KV-043-434

ORE GROUP C
TRES

Grower Talks®
on Plugs 3

Edited by

Jayne VanderVelde

WITHDRAWN

PERSHORE & HINDLIP
COLLEGE LIBRARY

12/00

CLASS	CODE
631.53	023157

Ball Publishing
Batavia, Illinois, USA.

023157

Ball Publishing
335 North River Street
Batavia, IL 60510, USA
www.ballpublishing.com

Copyright © 2000 Ball Publishing. All rights reserved.

Cover design by Tamra Bell.

No part of this book may be reproduced or transmitted in any form by any means, electronic or mechanical, including photocopying, recording, or any other storage and retrieval system, without permission in writing from Ball Publishing.

Reference in this publication to a trademark, proprietary product, or company name is intended for explicit description only and does not imply approval or recommendation to the exclusion of others that may be suitable.

Library of Congress Cataloging-in-Publication Data

GrowerTalks on plugs 3 / edited by Jayne VanderVelde.
    p. cm.
Over 40 articles from the pages of GrowerTalks magazine.
Includes index.
    ISBN 1-883052-24-6 (pbk. : alk. paper)
    1. Bedding plants—Propagation. 2. Plant plugs. 3. Bedding plant industry. I. Title: GrowerTalks on plugs 3. II. Title: GrowerTalks on plugs three. III. VanderVelde, Jayne, 1975-IV. GrowerTalks.
    SB423.7 .G76 2000
    635.9'62531—dc21
                                                    00-009228

Printed in the United States of America
05 04 03 02 01 00 1 2 3 4 5 6

LIBRARY
PERSHORE COLLEGE
AVONBANK
PERSHORE
WORCESTERSHIRE WR10 3JP

Contents

Contributing Authors

Teresa Aimone is a sales representative, Henry F. Mitchell Company, King of Prussia, Pennsylvania.

Mark Bennett is professor of horticulture, Department of Horticulture and Crop Science, The Ohio State University, Columbus, Ohio.

Chris Beytes is editor of *GrowerTalks* magazine, Batavia, Illinois.

Andrew Britten is head grower, Suncoast Greenhouses, Seffner, Florida.

Bruce Brodbeck is owner, Brodbeck Greenhouse, Toledo, Ohio.

Sherri Bruhn is a contributing editor for *GrowerTalks* magazine and editor of *Seed Trade News* magazine, Batavia, Illinois.

Dr. Daniel Cantliffe is professor and chairman, Horticultural Sciences Department, University of Florida, Gainesville, Florida.

Amy Compton is a graduate student, Department of Horticultural Science, North Carolina State University, Raleigh, North Carolina.

Brian Corr is New Crops and Development Manager, PanAmerican Seed, West Chicago, Illinois.

David Cross is a former production research specialist, PanAmerican Seed, West Chicago, Illinois.

Don Grey is a contributing writer for *GrowerTalks* magazine from Oregon City, Oregon.

P. Allen Hammer is professor of floriculture, Purdue University, West Lafayette, Indiana.

Dr. Mary Hausbeck is assistant professor and extension plant pathologist in vegetable and greenhouse ornamental crops, Department of Botany and Plant Pathology, Michigan State University, East Lansing, Michigan.

Will Healy is manager, Technical Services, Ball Horticultural Company, West Chicago, Illinois.

Jin-Sheng Huang is a graduate student, Department of Horticultural Science, Carolinia State University, Raleigh, North Carolina.

Philip Katz is cut flower product group leader, PanAmerican Seed, West Chicago, Illinois.

Dr. Peter Konjoian is president, Konjoian's Floriculture Education Services, Andover, Massachusetts.

Dr. Dave Koranski is with ETA Inc., Woodbury, Minnesota. He is a co-author of *Plug and Transplant Production, A Grower's Guide*, available from Ball Publishing.

Lisa Lacy is product manager, Novartis Seeds, Downers Grove, Illinois.

Miller B. McDonald is professor of seed physiology, Department of Horticulture and Crop Science, The Ohio State University, Columbus, Ohio.

Chris Millar is information systems manager, Knox Nursery Inc., Orlando, Florida.

Paul Nelson is a professor, Department of Horticultural Science, North Carolina State University, Raleigh, North Carolina.

Claudio C. Pasian is Floriculture Extension Specialist and associate professor, Department of Horticulture and Crop Science, The Ohio State University, Columbus, Ohio.

Joli A. Shaw is a former associate editor of *Grower Talks* magazine and former editor of *Green Profit* magazine, Batavia, Illinois.

Bill Sheldon is a freelance writer from Gardener, Kansas.

Dr. Roger C. Styer is president, Styer's Horticultural Consulting Inc. He is a co-author of *Plug and Transplant Production, A Grower's Guide*, available from Ball Publishing.

Paul and Paula Yantorno are part-owners of Center Greenhouses, Denver, Colorado.

Introduction

Plugs: Evolution and Revolution

In its brief history, the plug industry has undergone drastic changes. This field, not yet thirty years old, has progressed from a few growers experimenting with the tiny plants to an industry that produces billions of plugs each year for thousands of growers worldwide. And *GrowerTalks* magazine has been your eyes and ears on this booming trade since day one.

Growers now produce not only bedding plants from plugs, but also perennials. Tray manufacturers have collaborated to produce a common element (CE) tray for greater uniformity in equipment and automation, and greenhouse builders are designing ever-more efficient plug ranges. Billions of plugs per year are handled by machines that offer greater output, flexibility, and accuracy.

And the plug industry boom shows no signs of slowing down. With experts estimating that plug specialists sell only 30% of the plugs used in North America, the opportunity for even more growth is eminent. In fact, in our chapter on growers, we talk to some plug specialists who, in the wake of huge expansions, can hardly keep up with demand.

Continuing in the tradition of *GrowerTalks on Plugs* and *GrowerTalks on Plugs 2*, this latest compilation provides you with the most recent innovations in plug media, nutrition, and equipment from experts in the industry. Whether you're growing plugs yourself or buying them in, *GrowerTalks on Plugs 3* will inspire you with techniques and advice from successful plug growers, give you valuable cultural and technology details, and arm you with the information you need to become a more successful grower.

Chapter 1
Why Plugs?
General Information

The U.S. Plug Industry:
Bigger Plugs, Better Service

by Joli A. Shaw

Plug production in the last ten years has been marked by revolutionary advancements in seed germination and vigor along with stepped-up customer service. Unfortunately, plummeting prices and razor-thin margins have offset these improvements. Plug growers are meeting their customers' demands by growing larger plugs, using enhanced seed to get more usable plugs, and fine-tuning their variety lists.

We surveyed the country's top plug growers and finishers to get their views on today's plug industry. Here's what Bob Barnitz, Bob's Market and Greenhouse; Clay Murphy, Plants Inc.; Gerry Raker, C. Raker & Sons; Jack Van de Wetering, Ivy Acres; and Dennis Wenke, Wenke Greenhouses, see as the trends that are shaping plug production.

Lowering Costs with Higher Germination

"PreMagic (pregerminated seed) is doing a great job for us this year. It gives us a shorter time that the crop is in the greenhouse," says Jack Van de Wetering, Ivy Acres Inc., Calverton, New York. They haven't used the germination chamber this spring and have still achieved about 92% germination.

Cutting germination time means lower costs from production to shipping. "For example if the norm was thirty-five days, we can now do it in thirty days and take a percentage of time off the germination time," he says. Jack has seen improvements in his transplanters' efficiency and subsequently lower labor costs.

"It's possible with PreMagic to get one hundred usable seedlings routinely without a lot of effort," says Clay Murphy, Plants Inc., Huntsville, Texas. "As the seed companies learn more about seed and grading, from a preventative point of view, the seed going in will be much better than in years past."

But some growers still are not sold on the benefits of pregerm seed and have even reduced its use this year. "I see no need for pregermination except maybe for some seed producers to upgrade poor lots. An inconvenience we have found with the pregerminated seed is the need to use powder to dry it off—it sticks to the drum seeders," says Bob Barnitz, Bob's Market and Greenhouse, Mason, West Virginia.

Wenke Greenhouses, Kalamazoo, Michigan, starts with raw seed, then brings PreMagic impatiens into production when they get to "crunch time" to shorten germination two to three days. They used pregerminated seed for fall pansies last year, but will reduce the amount they use this year. They'll use more primed seed because contrary to their original plan, they still had to put pregerminated seed in the cooler. They couldn't put it on the greenhouse bench, so the benefit wasn't there, says Dennis Wenke.

Computer programs, like the Ball Vigor Index (BVI), that predict usable plugs have also helped growers meet the elusive goal of a 100% plug tray. "BVI probably precludes pregermination as long as they can keep up 90% or better," says Bob. "For two years now, I've averaged 92% on [impatiens] varieties like Dazzler, Super Elfin, Impulse, Showstopper, and Accent. That includes some that are not BVI like Accent Mix, Burgundy, and Violet, and some raw seed." His overall germination has increased 15% from two years ago, when he averaged 77% on impatiens.

These advancements also mean that growers have to meet their customers' higher standards. "It used to be that an 85% tray was good," says Clay. "Now a 95% tray is the norm."

Growers look to the day when 95% to 96% trays will be common, and they see it happening in the next five years. Price is no concern in this area. "It may cost more for seed, but growers are willing to pay for it—they'll see the benefit," Jack says.

Are 512s on the Way Out?

What is the biggest trend in plugs? According to these growers, it's larger sizes for quicker finishing at the end of the season. "We determined five or six years ago that we could cut seven days off our finish time on an impatiens flat by dropping from a 512 to a 406," Bob says. All of the trays he produces are 406 or larger. This season he's experimenting with 338 deep trays, which he says have soil volume equal to a standard 288. With the larger plugs, he's hoping to cut his production time by several more days.

Producing in larger trays such as 338, 288, and 144 allows growers to fill spot markets with quicker turns and more accurate finish dates. "We're using 512s and 216s because you have better predictions for when they'll be ready. With bigger plugs you can predict for certain markets," Jack says. Although producing fourteen-day bedding—getting four turns out of his operation—has been successful for Jack, he says it's not for everyone. "If you have the market, you can make it work. For example, if you have a market that wants impatiens every two weeks."

The concept isn't for every operation. "We're still in a 512 plug and have been for a while. We experimented at the end of the season with a larger plug. It's a space issue in the plug area for us—we don't see the dramatic difference between a 512 and, say, a 288, for example," says Dennis.

But are growers sacrificing quality for efficiency? Some growers doubt the ability to maintain product standards during such a fast production time. "I question some of the quality issues—whether you can get by with it," Clay says.

Competition Squeezes Margins

Increasing competition, a saturated market, and stagnant prices on finished bedding have taken their toll on margins, as growers are forced to take price cuts to meet the demands of mass marketers who want high-quality product at rock-bottom prices. "Our margins have dropped significantly. I think that will continue. It feeds on itself—if one [mass marketer] does it, it has a ripple effect," Jack says.

Like anything else, there is increased demand for higher quality at lower cost. "Our margins are squeezed, but tell me another industry in this economy that's different. That's a function of competition," Clay says.

Stagnant prices have spurred more growers to grow plugs. Five to ten years ago, few people produced plugs for sales. In October or September when growers are planning, it's hard for them to convince themselves they will make mistakes sowing, says Dennis. "Everybody has to use their greenhouses, particularly Northern growers, and plugs are a good way to fill greenhouses."

As more growers choose to grow their own plugs, overflowing markets force growers to hold plugs. This year's weather has worsened the situation. "Prices plummeted in the middle of April this year again," Dennis says. "We allow growers to keep their [purchased] plugs in the greenhouse for one week past ship date, and everyone's using it right up to the limit this year and asking for extensions."

Customer Service: Be Something to Someone

Mass marketers' entry into the market has forced growers to find niche markets and target their production. "In the past, brokers and producers could sell to all segments of the market. As it gets more competitive, different products zero-in on specific market segments. One segment supplies the mass markets, another segment supplies smaller customers and independents—the people who compete against mass markets," says Gerry Raker, C. Raker & Sons, Litchfield, Michigan.

Growers have adapted various strategies with their variety lists to compensate, some narrowing to supply only five or six varieties the mass marketers request, others widening to meet variety-specific demands of smaller customers and consumers. The argument continues: Should you be everything to everyone or something to someone? "From a production perspective you want [your variety list] as small as you can get it; from a marketing perspective you want it as wide as you can," says Clay.

For Jack Van de Wetering, the solution is simple: "Our variety list changes on a year-to-year basis—it's affected by pack trials, the fashion industry, the breeding companies. But in two to three years we have kept the same minimum order."

Adding to the confusion is the question of consumer knowledge. In reality, will end consumers prefer Showstopper or Dazzler impatiens? Or will they simply buy what's on the shelves? "We're narrowing our list. We base our selection on quality in the plug tray and what the customer wants," says Dennis Wenke. "The end user doesn't care whether he has Accent or Impulse—the mass market [buyer] cares more."

But growers serving small- and medium-sized customers disagree, saying they'd lose market share if they didn't offer more varieties. "We have sixty varieties of fall pansies—five to six series, over twenty impatiens and thirty petunias. It would hurt us if we didn't have the selection. We have no mass-market business," Bob says.

Customer service seems to be the key, and growers adjust their production accordingly, despite increased cardboard and paper prices and rising UPS and FedEx rates.

Though some have increased their minimum orders, little has changed in shipping and delivery. "We ship quality and service. It doesn't make sense if you grow good-quality plugs and you don't get them there because it costs too much," Gerry says.

"Our minimum guarantee of plugs in a tray is 92%. For some varieties and tray sizes, it's 96%," he says.

"We operate on the basis that we must have 100% plug sheets. Anything that helps us move toward 100% is great," says Clay.

Why Grow Your Own?

Growers say they save money by growing their own. Not only do they save money, they also guarantee quality plugs for finishing. And they only have one person to blame for poor rooting or double sowing. "I've brought in trays where I've never seen a single sown plug tray," Bob says. "We single sow all of ours. Two to three plants per cell is not good quality."

Growers who are selling plugs and producing finished bedding say growing their own plugs provides extra insurance for a quality product. "We don't do any patching. We bet that we'll be able to pick the best trays for our customers so we won't have to," says Dennis. "We sell the good trays to our customers and keep the poor germinating ones for our own production."

The Future

Growers see computers and electronic communication as two of the most important advances the plug industry has to embrace. Mass marketers are demanding electronic data interchange (EDI) for faster, more accurate transactions. Florists are designing Web sites and taking up to ⅓ of their gross from electronic sales. Growers will have to bring the technology into their sector of the industry and make a place in their budgets for electronic media.

"People are waiting for [technology] to happen, but the computer industry is going forward so fast that you just have to jump in and start learning,"

Gerry says. "People get used to being able to access information when they want it."

The information highway is only one part of plug production in the future. Plug production isn't just good growing anymore. To stay competitive and remain on top, plug growers will have to take advantage of new and existing technology to fight thin margins and rock-bottom prices and bring customer service to a new level.

June 1996.

Perennial Plugs—the Sensible Solution

by Paul and Paula Yantorno

With specialization a key to survival in the '90s, plug production can be one less specialty that growers need to worry about. And annual plugs aren't your only plug option. Buying in perennial plugs can make sense for growers who are looking at the expense, the time, and the space required to produce a quality perennial crop from start to finish. With top producers throughout the United States selling high-quality, ready-to-finish perennial plugs, there's no shortage of supply if you decide not to seed your own.

Under the right conditions, perennials grow fast. This gallon container finished in six weeks from one 70-size plug.
Photo: Paul and Paula Yantorno

Benefits of Plugs

Perennial seed germination can be very erratic. Starting a crop with plugs that are actively growing is one way to accurately predict a finished crop. For growers unfamiliar with perennial varieties, the guesswork has been taken out of what to grow. Many producers have regional collections or suggestions to fit into any marketing program. Variety lists that focus on the best and most important perennials have been identified and heavily researched.

Many growers who take into account the amount of space and handling required to get a crop to the plug stage will opt to let the plug producer do his part. Because many perennial varieties require a cold treatment before flowering, purchasing vernalized plugs saves time and money. One of the benefits of using vernalized plugs is having flowering perennial plants to sell the first year. Starting a perennial crop from plugs ensures a fast takeoff and quick turns from valuable greenhouse space. Finish time can be as short as six weeks for some varieties!

Liners are shipped ready to plant. The plug supplier has the responsibility of maintaining the inventory, and with accurate scheduling, the grower can have multiple ship dates and fresh perennial plugs to plant.

Perennial Plug Culture

Perennials are available in several plug sizes, such as 128-count (sold as 125s). Their finish time is slightly longer than 72-count (sold as 70s). These larger plugs are best when a quicker finish is required. They're good for a fast, four-inch finished pot size, and for wintering over. Plant one to three plugs per one-gallon container, depending on the variety and the plug size. For example, Blue Emerald *Phlox subulata* requires three plugs (either 125-count or 70-count size) per gallon, while one plug is recommended for Allegro *Papaver orientale*.

A soilless mix with good drainage and water retention is best for planting perennials. Perennials tend to be heavier feeders than foliage plants and trees or shrubs. Nutrition level recommendations include:

pH	5.5 to 6.8
Ammonia nitrogen	50 to 150 ppm
Nitrate nitrogen	350 to 800 ppm
Phosphate phosphorus	125 to 350 ppm
Potassium (potash)	350 to 800 ppm
Calcium	1,000 to 4,000 ppm
Specific conductance	150 to 225 mmhos

Perennial plugs shouldn't be held too long before planting. Keep them cool (45°F) and in bright light. Like annual plugs, perennial liners are moisture sensitive, so keep them from drying out.

Once planted, greenhouse temperatures (55° to 60°F nights and 60° to 65°F days) and the fertilizer regimen should be adjusted to promote root growth, while keeping shoot/crown growth as compact as possible. High-light levels and night lighting (400+ foot-candles) is also helpful in assuring good growth and compactness under winter's normally low-light conditions. As soon as the plant is well rooted (four to six weeks) and the crown is well-developed, greenhouse temperatures should be lowered to 35° to 40°F (day and night) for ten weeks for hardening and vernalization.

Vernalization

Generally, 70-count and 125-count plugs, purchased non-vernalized, can be planted in one-gallon pots in January after Christmas crops have shipped. Hardened off from February 15 through March 1 at 32°F, they can then be vernalized outside (freeing up greenhouse space for bedding annuals) from March 1 through shipping. Plant 70-count plugs in four-inch pots during week two or three; harden off at 32°F from February 15 through March 1; vernalize outside from March 1 through shipping. If outside temperatures are going to be below 30°F, newly planted plugs should be covered at night, or grown inside for at least two weeks or until rooted. After two weeks, perennial plugs can handle cooler temperatures.

Michigan State University research has found that juvenility is one factor in successful vernalization. Cold treatment is effective only if the plug is old enough. Plug size doesn't always mean older plugs. Although it differs by variety, ten to sixteen leaf pairs seem to be enough for the vernalization process. Good plug producers use this as a measure of readiness.

Some perennial varieties that benefit from vernalization, such as *Lavandula angustifolia* and *Liatris spicata*, can be purchased pre-vernalized, which can cut your finishing time. Many perennials, especially mid-summer through fall blooming varieties and many ground covers, don't require vernalization, so they can be scheduled for later planting, similar to other flowering crops. Growing on at cold temperatures is beneficial for producing a high-quality finished perennial, both inside and outside the greenhouse.

Fall Sales and Overwintering

Summer- or fall-planted perennial crops can be used for fall sales and to winter over. Production can be handled exclusively outdoors, or inside the greenhouse, depending on the type of facility available, the climate, and the size liner purchased. Slower growing varieties that rely on underground buds or roots for survival and 125-count plugs, should be planted in late July or early August to allow adequate rooting and bud development before the onset of severe weather. Faster growing varieties in 125-count and 70-count plugs can be planted during August or September, depending on species. Keep fertility on the higher end of recommended ammonium-nitrogen rates and moderate nitrate-nitrogen rates initially to promote stem growth, branching, and bud set. To help harden stem growth and promote root development, decrease the ammonium nitrogen as the fall season progresses.

A bushy, well-rooted plant with a well-developed crown will survive the next steps of drying down and mild stressing, necessary to further the hardening process. Once the plants appear dormant and soil in the container is frozen, provide winter protection. Prior to covering with a frost cloth, check for excessive moisture and use a cover spray of fungicide to prevent crown rot and foliar diseases. Winter watering is necessary in cold, dry climates like Colorado.

Uncover perennials promptly in the spring to prevent stretching. A spring fertilizing plan with moderate levels of nitrate nitrogen will assure a harder perennial with a healthy flush of spring growth.

Pest Control

Perennials are susceptible to many of the same insect and disease problems as other plants. Good perennial plug suppliers will have screened stock and propagation areas and will ship only clean cuttings. Watering early in the day and as infrequently as possible, combined with proper pot spacing and good ventilation, will reduce the chance for disease contamination. Using a fertilizer high in nitrate nitrogen rather than ammonium nitrogen will also help produce a healthy perennial crop.

The control of powdery mildew, botrytis, and rusts are a special concern in winter and early spring. Preventative measures such as sulfur burners and fungicide cover sprays are essential to successfully grow perennials such as veronica, verbena, and monarda.

Monitoring insects with sticky traps, then applying pesticides or biological controls as needed, will help prevent serious insect infestations. Thrips carry tomato spotted wilt virus and impatiens necrotic spot virus, which are easily spread and symptomatic only at high temperatures (over 70° to 75°F). Perennials are asymptomatic—they don't show symptoms of these viruses, but they can be carriers. Purchasing liners from a reputable supplier ensures clean plants. Prevent foliar diseases with fungicide cover sprays such as Clearys, Phyton, Subdue, and Benomyl. Consult your local extension office for assistance and specific pest problems.

Why Do It All?

Perennial plug producers have a lot to offer. They can help you focus on your specific area of expertise—that of providing a top-quality finished plant. Buying in perennial plugs can save time, money, and valuable greenhouse production space. Purchasing perennial plugs is the sensible answer for the finish grower who asks the question, "Should I try to do it all?"

February 1997.

Chapter 2
The Proper Plug Diet: Media, Fertilizer, & Nutrition

[decorative symbol]

Key Factors of Water, Media, and Nutrition

by Dr. Roger C. Styer

Ever wonder why your plugs turn yellow or you don't get the white, fuzzy roots everyone talks about? Do you have trouble controlling shoot growth regardless of the weather? These are common problems for plug growers. Sometimes growers will use a different media or fertilizer and eliminate their previous problem but create a new one. Understanding the key factors of water and media quality, nutrition, and how these affect the growth of your plugs is essential for solving common plug problems and consistently growing quality plugs.

About 75 to 80% of all nutritional problems are due to media pH and soluble salts. How do you measure these two factors on a consistent basis? If you do measure media pH and soluble salts, what levels should you maintain and why? What type of fertilizer should you use to control plug growth and why?

Water Quality

Alkalinity

The most important factor to know about your water is alkalinity—water's capacity to neutralize acids (H+). Think of it as the buffer capacity of water. The alkalinity level is determined by the total amount of dissolved bicarbonates, carbonates, and hydroxides. Plug growers can measure alkalinity by using a simple test kit or by sending a water sample to a testing lab.

Another way of thinking about alkalinity is that it's like lime in the water. The higher the alkalinity levels, the more rapidly media pH rises. When media pH rises above 6.5 for most crops, uptake of most minor elements, such as iron and boron, is reduced, resulting in minor element deficiencies. This reaction occurs quickly in a plug cell but much slower in larger containers.

11

Sometimes alkalinity can be too low, and the water has no buffering capacity. This means media pH will fluctuate more, depending on what type of fertilizer you use.

The optimum range for alkalinity for plugs is 60 to 80 ppm. If your water contains higher levels than desired, you can neutralize the excessive alkalinity by injecting acids such as sulfuric, phosphoric, nitric, or organic. Water alkalinity levels greater than 350 ppm will need treatment systems such as reverse osmosis or an alternate water source.

Soluble salts

Soluble salt, or EC, is a measure of all dissolved salts in the water and includes nutrients (nitrogen, calcium, magnesium) and non-nutrients (sodium, chloride). Soluble salts are measured with a conductivity meter and quantified as millimhos per centimeter (mmhos/cm) or millisiemens per centimeter (mS/cm). Generally, soluble-salt levels for irrigation of plugs should be less than 1.0 mmhos/cm.

High concentrations of soluble salts can decrease germination, damage roots and root hairs, and burn leaves of young seedlings. Damage will differ depending on the crop. You need to know which salts make up high soluble-salt levels in the water. The most common salt problems include sodium, chloride, boron, fluoride, and sulfate. If these levels aren't too high, then leaching with each watering and properly drying media between watering may be sufficient. However, very high levels of sodium or chloride will require reverse osmosis water treatment or an alternate water source.

Sodium absorption ratio

The sodium absorption ratio quantifies sodium levels in relation to calcium and magnesium levels in the water. If the sodium absorption ratio is less than

2.0 and the sodium level is less than 40 ppm, then you should have no problem with sodium from the water. High sodium levels will cause media to tighten up, holding more water and less air. This will reduce root growth significantly. Sodium competes with calcium, magnesium, and potassium. Use fertilizers with more calcium, magnesium, and potassium, starting early in the plug cycle. Be sure to leach 5 to 10% with every watering to remove excess sodium.

Table 1. Water Quality Guidelines for Plug Production	
pH	5.5 to 6.5
Alkalinity	60 to 80 ppm CaCO₃
Soluble salts (EC)	less than 1.0 mmhos/cm
Sodium Absorption Ratio (SAR)	less than 2
Nitrate (NO₃)	less than 5 ppm*
Phosphorus (P)	less than 5 ppm
Potassium (K)	less than 10 ppm
Calcium (Ca)	40 to 120 ppm
Magnesium (Mg)	6 to 25 ppm
Sodium (Na)	less than 40 ppm
Chloride (Cl)	less than 80 ppm
Sulfate (SO₄)	24 to 240 ppm
Boron (B)	less than 0.5 ppm
Fluoride (F)	less than 1 ppm
Iron (Fe)	less than 5 ppm
Manganese (Mn)	less than 2 ppm
Zinc (Zn)	less than 5 ppm
Copper (Cu)	less than 0.2 ppm
Molybdenum (Mo)	less than 0.02 ppm

*ppm = mg/l

The pH of the water does not directly affect the pH of the plug media; the alkalinity does. Keep water pH between 5.0 and 6.5 for best solubility of growth regulators, fungicides, and pesticides. High levels of boron, chloride, and fluoride can also present problems (Table 1).

Media Quality

The two main physical characteristics to know about any plug media are water-holding capacity and air porosity. The water-holding capacity of media affects the frequency of watering. Capillary pore spaces (very small pores between particles) will hold water against gravity, thereby contributing to water-holding capacity. The finer the particle sizes in the plug media, the greater the water-holding capacity.

The opposite of water-holding capacity is air porosity, which is determined by the amount of noncapillary pore space in the media. Noncapillary pore spaces are larger pores between particles. Water drains from them by gravity, and air moves in through them. In a plug cell, the air porosity may range from less than 1% up to 10%. Air porosity can be reduced in plug media with fine particles in smaller plug cell sizes and through media handling. Sufficient air porosity is needed for good root growth, drying out quickly, and controlling algae and shoot growth.

Due to the small size of plug cells, the chemical properties of media change rapidly depending on your water quality, your growing environment, and the way you water and fertilize. Whether you mix your own plug media or buy a commercial mix, you need to know what the starting pH is, how it will change in the first two weeks, how much soluble salt (EC) is present, and what nutrients are available to the seedling in the first two weeks.

Media pH

The media pH controls the availability and uptake of all essential plant nutrients. Generally, with a 5.5 to 6.5 pH for most crops, all nutrients are readily available in a peat-based media. Below a 5.5. pH, major elements, such as calcium and magnesium, become unavailable, while minor elements become too available. Above a 6.5 pH, minor elements become deficient, while calcium is oversupplied.

To make up for the acidic nature of peat-based plug mixes, pulverized limestone is used to raise the pH level. How fast the media pH goes up in the first two weeks depends on the buffering capacity (ability to resist change) of the growing media; the amount, mesh size, and type of limestone used; how much moisture the plug media had during storage; and the watering frequency during germination (more water speeds up lime release). Remember water alkalinity will also influence media pH.

You need to test your plug media straight from the bag or immediately after you make it, as well as during the first two weeks of the plug cycle. Fill trays with the media, water them with distilled or deionized water as you normally would for the first two weeks, but don't put seed or fertilizer in trays. Test the media pH weekly. If the pH goes up too rapidly, then limestone is the cause.

EC

High soluble salts (greater than 1.0 mmhos/cm, 2:1 dilution) in the plug media to begin with can reduce germination and burn-off sensitive primary roots of emerging seedlings in crops such as vinca, snapdragon, impatiens, and begonia. If your EC reading is high in the first two weeks and you haven't added more fertilizer, then your starter fertilizer charge is the problem. Most commercial plug mixes now have fairly low starter charges. Know what nutrients make up the soluble salts in your starting plug media. If your starter charge is too high in ammonium, then germination and initial root growth for many crops will be reduced. If it's very low, then you may need to start feeding some crops earlier than expected. For most crops and conditions, the starter charge will last for the first seven to ten days after seeding.

Not all commercial plug mixes are the same, nor are all batches of plug mix from the same company exactly the same. Make sure you test your plug media every year and in every load you receive (Table 2).

Nutrition

You should know three key factors about any fertilizer to understand what it will do for the plug. On a commercial fertilizer, these three key

Table 2. Acceptable Initial Plug Media Nutrient Levels	
pH	5.5 to 5.8 for most crops
	6.0 to 6.2 for low pH-sensitive crops
EC mmhos/cm	0.4 to 1.0 depending on the crop
Nitrate (NO₃)	40 to 60 ppm*
Ammonium (NH₄)	less than 10 ppm
Phosphorus (P)	5 to 8 ppm
Potassium (K)	50 to 100 ppm
Calcium (Ca)	60 to 120 ppm
Magnesium (Mg)	30 to 60 ppm
Sulfur (S)	50 to 200 ppm
Sodium (Na)	less than 30 ppm
Chloride (Cl)	less than 40 ppm
Boron (B)	0.2 to 0.5 ppm
Iron (Fe)	0.06 to 6 ppm
Manganese (Mn)	0.03 to 3 ppm
Zinc (Zn)	0.001 to 0.6 ppm
Copper (Cu)	0.001 to 0.6 ppm
Molybdenum (Mo)	0.02 to 0.15 ppm

Values based on saturated media extract (SME) procedure.
*ppm = mg/l

factors are on the label. They include percentage of total nitrogen as ammonium and urea, availability of calcium and magnesium, and how acidic or basic the fertilizer is.

Basically, there are three types of nitrogen—nitrate nitrogen (NO_3), ammoniacal nitrogen (NH_4), and urea nitrogen—and each has a different effect on plug growth. Nitrate nitrogen is most available to the plant. Ammoniacal nitrogen must be broken down by soil bacteria to become available to the plant in the form of nitrate nitrogen. This bacterial conversion slows down greatly below 60°F, resulting in ammonium toxicity. Urea nitrogen must be converted to ammoniacal nitrogen first, then to nitrate nitrogen to be used by the plant. For this reason, we group urea nitrogen with ammoniacal nitrogen.

Fertilizers containing more than 25% ammoniacal nitrogen will promote larger and softer amounts of shoot growth but not root growth. Fertilizers

Table 3. Common Commercial Plug Fertilizers[1]

Fertilizer	Ammoniacal nitrogen %[2]	Potential acidity[3]	Potential basicity[4]	Calcium percentage[5]	Magnesium percentage[5]
21-7-7	100	1,560		—	—
9-45-15	100	940		—	—
20-20-20	69	583		—	—
20-10-20	40	422		—	—
21-5-20 (Excel)	40	418		—	—
15-15-15*	52	261		—	—
15-16-17*	30	165		—	—
20-0-20	25	40		5	—
17-5-17	24	0	0	3	1
17-0-17	20		75	4	2
15-5-15 (Excel)	28		135	5	2
13-2-13	11		200	6	3
14-0-14	8		220	6	3
15-0-15	13		420	11	—

[1]List of some commercially available fertilizers used for plugs. Not all formulations are the same from every company. Check the label.
[2]Ammoniacal nitrogen % is the total nitrogen percentage that's in the ammonium plus urea forms; the remaining nitrogen is nitrate.
[3]Pounds of calcium-carbonate limestone required to neutralize the acidity caused by using one ton of the specified fertilizer.
[4]Application of one ton of the specified fertilizer is equivalent to applying this many pounds of calcium-carbonate limestone.
[5]Only where percentage of calcium or percentage of magnesium were 1% or greater.
*Contains sodium nitrate (nitrate of soda), which adds unwanted sodium to plugs.

containing more than 75% nitrate nitrogen will promote root growth over shoot growth, resulting in shorter internodes and smaller, lighter green leaves but thicker leaves and stems (Table 3). Commercial fertilizers vary widely in the percentage of total nitrogen in the ammonium form. You can calculate this percentage from the information on the label by adding the percentage of ammonium and the percentage of urea, then dividing by the total percentage of nitrogen, and multiplying by 100. Keep these percentages in mind when selecting a fertilizer for plug growth. Do you want to push the plugs (speed up shoot growth), or do you want to slow the crop down and tone it up (encourage root growth and thicker leaves and stems)?

Calcium and magnesium are of primary importance to plug growers. As seen in Table 3, many fertilizers don't contain appreciable amounts of these two elements, especially fertilizers high in ammoniacal nitrogen. Calcium plays an important role in cell walls and root growth. Magnesium is needed for overall green color of leaves and good photosynthetic activity.

Most calcium and magnesium can be supplied in media with dolomitic limestone. Additional amounts may come from the water source. However, dolomitic limestone is activated within the first two weeks in a plug crop and may run out for longer term crops. Also, plug media manufacturers have been reducing the amount of limestone added in order to have the plug media start and stay at a lower pH (5.5 to 5.8). Therefore, it's important for plug growers to use some fertilizers containing calcium and magnesium in their feed program to supplement other sources and help control plug growth. Try to maintain a 2:1 calcium-to-magnesium ratio at all times.

Fertilizers high in ammoniacal nitrogen will cause an acidic reaction in the soil solution and will bring media pH down over time. Fertilizers high in nitrate nitrogen will cause a moderate alkaline reaction in the soil solution and will slowly raise media pH over time. To check the potential acidity or basicity of a commercial fertilizer, look on the label. Table 3 shows the potential acidity or basicity of common commercial plug fertilizers. The higher the number, the more acid or basic the fertilizer will be. For example, 20-10-20 has a potential acidity of 422, and 20-0-20 has a potential acidity of only 40. This means that 20-0-20 will change the media pH less compared to 20-10-20.

Winter 1996.

Plug Seedling Soil Sampling—Timing Is Critical

by Amy Compton and Paul Nelson

While visual and foliar diagnosis have their places in finding and diagnosing plug nutritional disorders, the best method of early detection is soil (root media) testing. Media tests forecast future problems better because quantities of nutrients present in the media that have not yet been taken up by the plug are measured. Problems can be detected at their inception, if regular media testing is practiced.

Fairly reasonable standards exist for media pH and EC levels for plug crops. However, specific nutrient standards have not yet been developed. Commercial labs have some tentative standards, and Michigan State University suggests using the "Acceptable" category within their greenhouse media standards table. This is one level down from their "Optimum" category for crops in general. These values are presented in Table 1. Plug seedling growers have attempted to use media testing, but many have found that the test results for specific nutrients are too variable and have abandoned media testing.

Table 1. Acceptable Media Nutrient Levels for Plug Seedling Crops

These encompass mainly the "Acceptable" range from the Michigan State University standards table for greenhouse media extracted by the saturated media extract procedure (saturated paste method)*.

EC (dS/m)	0.75 to 2.0	potassium	60 to 150 ppm
nitrate-N	40 to 100 ppm	calcium	80 to 200 ppm
ammoniacal-N	less than or equal to 20 ppm	magnesium	30 to 70 ppm
phosphate-P	3 to 5 ppm		

*All but ammonium from Warncke and Krauskopf 1983.

We suspected this variation was due to the difference in time within fertilization cycles when growers were sampling media. If the sample was taken after a watering versus after a fertilization, much lower concentrations could be expected. When a tray of large, rapidly growing seedlings was sampled sixteen hours versus two hours after a fertilizer application, a low test result could be expected. This seemed plausible because seedling growers have speculated that plants can deplete nutrients in a plug cell between fertilizations. Based on these suspicions, we developed the hypothesis that media test variation is due mainly to the dilution and leaching effects of water application as well as to plant uptake of nutrients over time.

In order to test our hypothesis, we grew Primetime White petunia seedlings in 288-cell plug trays in a 75% sphagnum peat moss/25% perlite media in two experiments. We used six fertilization/irrigation regimes including all combinations of two leaching percentages (0% or 20% excess fluid at each irrigation and fertilization) and three fertilization frequencies (each irrigation, every second irrigation, or every third irrigation). Fertilizer concentrations were adjusted to compensate for leaching percentage and irrigation frequency. Seeds were sown on October 19, 1993, and April 17, 1994; germinated in a dark chamber at 75°F; then grown in a glass greenhouse. Media solution was sampled by squeezing it out of the media through cheesecloth. When we sampled eight hours or longer after a watering or fertilization, we applied water one hour before sampling so there was no leaching.

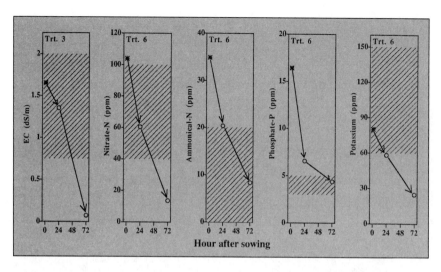

Figure 1. Shifts in soluble salt, nitrate-N, ammoniacal-N, phosphate-P, and potassium concentrations in root media within a single fertilizer cycle. Solid symbols denote media concentrations one hour after fertilizer application, while open symbols show concentrations one hour after each water application following fertilizer application. The shaded zone represents acceptable levels of EC or nutrients.

Nutrient levels in the media solution were considerably lower after a watering than after a fertilization (Figure 1). The differences were so great that opposite interpretations of the status of a fertilization program could be made within a single fertilization cycle. The situations depicted in Figure 1 are typical of the results for each nutrient within each treatment in both experiments. In the figure the soluble-salt (EC) level was at a desirable level one hour after fertilization but was deficient twenty-four hours later, one hour after a subsequent watering.

The total soluble-salt level is a good predictor of the condition of all other nutrients because generally the main contributor to soluble salts in the media solution is fertilizer. Nitrate levels in the media solution presented an even worse situation. One hour after fertilization the nitrate level was slightly excessive, yet the morning after a watering the level had dropped all the way through the adequate range to a deficient level.

Depending when the media sample was taken within this single fertilization cycle, one could draw the conclusion that the fertilizer concentration should be reduced or increased. Ammonium and phosphate levels decreased from unacceptably high to adequate, and potassium decreased from acceptable to deficient after a watering. Thus, dilution and leaching effects of watering dictate that media samples be taken after one or the other of these events but not randomly after either. We suggest that samples be taken after fertilization.

Nutrient uptake effects with passing time were likewise devastating to one's ability to interpret media analyses (Figure 2). On the sixteenth day after

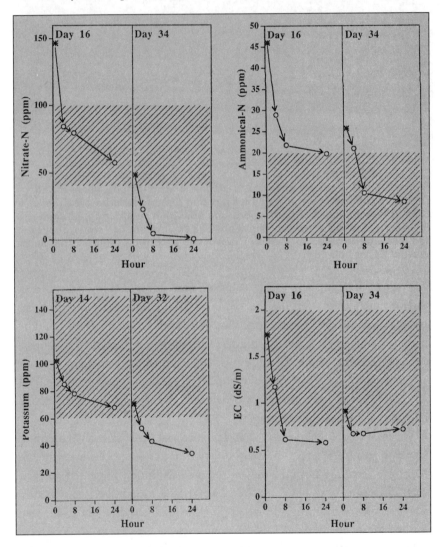

Figure 2. Soluble salts (EC), nitrate-N, ammoniacal-N, and potassium concentrations in root media one, four, eight, and twenty-four hours after two fertilizer applications: one during early production and the other during late production of a petunia plug crop. The shaded zone represents acceptable levels of EC or nutrients.

sowing, the total salt level fell from an adequate to a deficient level in eight hours after fertilization. We applied no water during that eight hours. On the thirty-fourth day of the crop, the salt level dropped from adequate to deficient in four hours. Similar shifts can be seen for nitrate, ammonium, and potassium in Figure 2. In each case media concentrations dropped from either excessive to adequate or adequate to deficient in four to eight hours.

Not only is it important that media be sampled after a fertilization as opposed to a watering but that it also be sampled within a specified time after fertilization. This becomes even more important as the seedling grows larger and depletes the media faster. It is necessary that you sample at least one hour after fertilization to give the newly applied fertilizer solution time to equilibrate with the media solution. On the other hand, don't wait too long, or plant nutrient uptake will change media solution concentrations. A period of one to two hours after fertilization would be best.

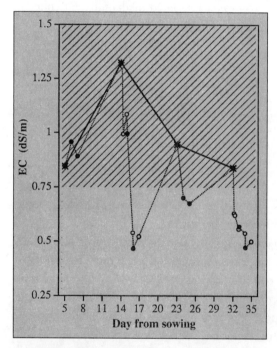

The effects of watering plus nutrient uptake over time on the media soluble-salt level can be seen in Figure 3. Fertilizer was applied at every third irrigation. The dotted curve connecting salt levels one hour after fertilizations and waterings plus various other times after fertilization on the fourteenth and thirty-second days is very erratic. It repeatedly crosses from adequate to deficient zones, leading to conflicting conclusions about the fertilizer program. When

Figure 3. Media concentrations of EC one hour after fertilization (stars); one hour after watering (solid circles); and four, eight, twenty-four, forty-eight, and seventy-two hours after fertilization on the fourteenth and thirty-second days (open circles). All sample times are connected by the dotted line, while only samples taken one hour after a fertilization are connected by the solid line. The shaded zone represents acceptable EC levels.

only salt concentrations one hour after fertilizations are connected, a smooth interpretable curve results. Salt concentrations along this curve rise at first because fertilizer application rate exceeds plant uptake. Later, concentrations fall because uptake by larger plants exceeds fertilizer delivery rate. Growers sampling this crop only after fertilizations could effectively use media testing, while growers sampling at various times along the dotted curve couldn't.

Media sampling can be very effective for monitoring and assessing plug seedlings' nutrient status. However, it's imperative that you take samples only after fertilizations and only at a set time one to two hours after fertilizations. Until more refined standards can be developed for plug seedling crops it's possible to continue using current media standards. By carefully relating the condition of each crop to its associated media nutrient concentrations, growers can personally refine these standards.

March 1996.

Media Testing: Talk to Your Lab

by Dr. Peter Konjoian

You've got many choices when it comes to selecting a lab to test your growing media for nutrients and soluble-salt levels. But are the results you get accurate and consistent? Is communication important with the people conducting the tests and interpreting the results?

We conducted an experiment to see how five different testing labs would handle our media and what kinds of readings we'd get back. The results of our test illustrate that good communication between you, your media testing lab, and the people interpreting the test results is critical if the tests are to help you grow better crops.

Historically, land-grant universities have performed the lion's share of soil testing, both for field agriculture and greenhouse crops. Now, however, the science of growing media has advanced to the point where most media manufacturers have soil testing labs of their own, staffed with Ph.D.-level experts. This has helped take up some of the slack created by budget cuts that severely impacted state- and federal-funded soil testing labs. University labs, particularly those in smaller states, have been forced to concentrate on servicing the field crops segment of agriculture because more of their samples originate in this sector. The private labs specialize in servicing the soilless media segment—you!

Testing the Testers

The extraction procedure for soil-based growing mixes differs from that for soilless mixes and is based on using acid to extract nutrients from the sample. Soilless media require a milder extraction procedure due to their dominant organic base. Smaller labs don't always have enough soilless media samples to warrant separate procedures and may opt to send soilless samples through the acid extraction procedure. When this happens, the test may produce readings that are artificially inflated for certain nutrients. If interpreted incorrectly you can be misled by the numbers.

The experiment

We designed a test in cooperation with five labs to generate data that would address this situation. Three of the labs are private and two are land-grant university labs. One of the state university labs is large and services growers both within and outside its borders; the other is small and services growers predominantly within its state. Additionally, the small lab tests more field soil samples than greenhouse soilless media samples. For this experiment the small lab ran the samples through both the soil-based and soilless extraction procedures.

Materials and methods

Unrooted poinsettia cuttings were stuck on July 5, pinched on August 29, and planted into five-inch azalea pots on September 14. Three growing media were used to simulate a range of mixes used by commercial growers. Metro Mix 510 was used as a representative soilless medium, a 15% topsoil mix represented a transitional mix, and a 30% topsoil mix represented a field soil-based mix. A second Metro Mix 510 sample incorporated slow-release Osmocote prior to planting.

A second treatment series included different rates of liquid fertilization on a constant feed basis. Peters Excell Cal-Mag (15-5-15) was used at the rates of 100, 250, and 400 ppm nitrogen. Media samples were collected from each of the four mixes at planting and from each of the twelve treatments (four mixes with three different fertilizer levels) four, eight, and twelve weeks after planting. There were 180 plants in the experiment which allowed for five plants to be sampled from each treatment on each sampling date.

Finally, plant height and width were measured at the experiment's completion as indicators of commercial quality and to benchmark the lab findings.

What We Found

All experimental treatments produced commercially acceptable plants, which helps us evaluate the lab test results. Table 1 presents height and width of plants grown in each of the four mixes averaged

Table 1. Effect of Different Media on Plant Size

| | Final measurements | |
Medium	Height (in.)	Width (in.)
Metro 350 with Osmocote	11.6	14.4
Metro 350 without Osmocote	11.7	13.8
15% soil	12.9	16.0
30% soil	12.8	16.1

over the three fertilizer rates. Plants grown in soil-based mixes were slightly larger in height and width than those grown in soilless mixes.

Table 2 presents final height and width for the three fertilizer rates averaged over the four mixes. The largest plants were produced with a 100 ppm rate, and the smallest plants resulted from the higher than usual rate of 400 ppm.

Table 2. Effect of Different Fertilizer Regimes on Plant Size

| | Final measurements | |
Fertilizer	Height (in.)	Width (in.)
100 ppm	13.1	16.9
250 ppm	12.1	15.4
400 ppm	11.5	13.0

As an aside, individual treatments showed that Osmocote incorporation into Metro Mix 510 increased plant size when used in conjunction with 100 ppm liquid feed, had no effect with 250 ppm, and inhibited plant size when used in conjunction with 400 ppm liquid feed. Also, plants in soil grown at 100 ppm of liquid feed were larger than those grown in the soilless mix, unless slow-release fertilizer was incorporated.

The Lab Results

pH

The labs were randomly coded A through E. Lab E was the smaller university lab that provided analyses using both field soil and soilless media extraction methods. Table 3 presents data from two of the twelve treatments: Metro Mix 510 without Osmocote plus 250 ppm liquid fertilization and 15% soil plus 250 ppm liquid fertilization.

Table 3. Test Results: Media pH

Lab	Metro Mix 510 without Osmocote, 250 ppm nitrogen			15% soil, 250 ppm nitrogen		
	Week 0	Week 12	Change	Week 0	Week 12	Change
A	6.06	5.14	-0.92	6.41	6.14	-0.27
B	5.50	4.90	-0.60	5.50	5.70	+0.20
C	5.50	4.80	-0.70	5.80	5.50	-0.30
D	5.80	4.70	-1.10	5.80	5.70	-0.10
E soilless	5.80	4.90	-0.90	5.70	5.60	-0.10
E soil	5.60	5.10	-0.50	5.50	5.60	+0.10

Lab A consistently reported the highest pH values over the entire experiment. The pH change from week zero to week twelve was negative (drop in pH) in ten out of twelve cases and positive (increase in pH) in the other two. The magnitude of the change was greater in the soilless mix than in the 15% soil mix, as we'd expect due to the additional buffering capacity of the soil. Differences in pH measurement due to Lab E's two extraction procedures were minimal.

Soluble salts

Table 4 presents soluble-salt readings for Metro Mix 510 at the start of the experiment (week zero) and at the end (week twelve) both with and without Osmocote. Beginning readings without Osmocote were similar for all five labs. However, incorporating Osmocote did affect readings. The procedure for testing media with incorporated slow-release fertilizer was to remove the fertilizer "prills" from the mix by hand prior to testing.

Table 4. Growing Media Soluble-Salt Levels (mmhos)

Metro Mix 510, 250 ppm nitrogen

Lab	Week 0		Week 12	
	Without Osmocote	With Osmocote	Without Osmocote	With Osmocote
A	0.98	1.55	2.04	3.42
B	0.78	2.16	2.26	3.95
C	1.00	2.30	2.80	3.50
D	0.76	2.55	3.06	4.11
E soilless	1.08	3.48	3.16	5.79
E soil	0.55	1.39	1.83	2.48

LIBRARY
PERSHORE COLLEGE
AVONBANK
PERSHORE
WORCESTERSHIRE WR10 3JP

The highest soluble-salt readings at week zero were obtained by Lab E when using its soilless extraction method. Because Lab E's reading of 3.48 is about one full point above the rest of the readings, it's possible that Lab E did not remove prills as thoroughly as the other labs. Remember that this sampling was right out of the bag with the Osmocote incorporated just prior to mailing samples. No moisture was added to the mixes, and results were received from all five labs within a period of one week, verifying that each of the labs processed the samples in a similar, timely fashion.

Most labs note an acceptable range for soluble salts to be 0.75–3.5 mmhos. Lab E's reading of 3.48 could be interpreted to mean that the mix has a soluble-salt problem, raising concerns for the grower that aren't real. At the completion of the experiment, Lab E's soilless reading was again high compared to the other labs. The 5.79 is off any of the charts, but Table 1 shows that this treatment produced plants of similar quality to other treatments. Without this proof that the plants grew well, this soluble-salt reading could be mistakenly interpreted as a severe problem. Lastly, it's interesting to note that Lab E's soil procedure yielded consistently lower soluble-salt readings than any other lab.

Individual nutrients

Levels of selected, individual nutrients at the completion of the experiment are presented in Table 5 and include phosphorous, potassium, calcium, and magnesium. The treatment presented is that of Metro Mix 510, without Osmocote, fertilized with 100 ppm nitrogen. The differences seen in this treatment are typical of those observed throughout the experiment.

Table 5. Growing Media Nutrient Levels

Metro Mix 510, 100 ppm nitrogen, Week 12 (ppm)

Lab	Phosphorous	Potassium	Calcium	Magnesium
A	5	118	29	16
B	2	90	23	17
C	5	121	30	17
D	6	127	31	17
E soilless	8	158	42	22
E soil	21	457	1,864	420
Normal range	5 to 50	35 to 300	40 to 200	20 to 100

The most obvious difference is seen in the results from Lab E, which ran the sample through an acid extraction procedure as if it were a soil-based sample. In the case of each nutrient in the table, the figures were considerably higher than any other lab's figures. Phosphorous levels were more than three times higher than any other lab's and more than twice as high as Lab E's level for the soilless procedure. Potassium was over three times higher than the other labs and just under three times higher than the soilless procedure. Calcium and magnesium levels reported from the soil procedure were over forty and twenty times higher, respectively, for the soil procedure than for any of the soilless extractions. Remember, calcium and magnesium are problem nutrients for poinsettias and require frequent monitoring. Additionally, note the bottom row of the table that presents accepted ranges for normal growth and development. If interpreted incorrectly by the person making a recommendation or by the grower who receives the results, the conclusion could be that there's plenty of calcium and magnesium available to the crop. In reality the other four labs as well as Lab E's soilless procedure all agree that both calcium and magnesium levels fall just below the low end of the acceptable range. That's a swing in possible interpretation from slightly deficient to strongly excessive. Again, keep in mind that the plants grew normally.

What It All Means

Two key points were supported by this experiment. First, on the negative side, running a soilless sample through a soil-based extraction procedure yielded different readings than a traditional soilless extraction procedure. Second, but a positive finding, four of the five labs participating in the experiment produced very consistent readings.

You shouldn't jump from lab to lab. Pick one and establish a regular pattern of nutrition monitoring. The differences between labs aren't as important as the consistency from sample to sample and crop to crop.

It's not the intent of this article to be critical of Lab E's testing procedures. Instead, the intent is to point out that careful communication between grower and lab as well as between lab and interpreter is very important. In several instances data generated in this experiment could have been misinterpreted and caused unnecessary concern for the grower. This communication is a responsibility that should be shared by all parties involved: media manufacturer, grower, lab scientist, and extension specialist.

July 1997.

Chapter 3
Seed, Germination, and Growing

Seed Viability, Germination, and Vigor: Sorting out the Terminology

by Miller B. McDonald

Rapid, uniform seed germination is an important, but often neglected, factor in profitable bedding plant production. Plug producers recognize that reductions in germination and the time necessary for germination and seedling establishment lower costs. To help plug producers achieve this goal, many flower seed companies focus on seed enhancements such as priming, pelleting, and pregermination.

Here's an example of the significance of seed quality to the flower plug producer: A grower purchases a packet of impatiens seed with a reported 95% germination. In a 512-plug tray, the grower determines that 80% of the seed planted produced seedlings in each cell. At a market cost of $.03 to $.05 per cell, this is a $3 to $5 loss per tray for the grower. Also, the grower isn't using 20% of his greenhouse space to produce plants.

Even though seed quality is very important, terms and concepts critical to seed quality are often confused. Here we help you define and clarify seed viability, germination, and vigor for plug production.

Viability

Seed viability and germination are often considered synonymous terms, but this isn't correct. Seed viability means the degree to which a seed is alive. In the above example, why wasn't 100% germination achieved in this impatiens seed lot? It may be that 5% of the remaining seeds were dormant. If this is so, they aren't dead and have the capability to germinate at some point in time, if dormancy is broken. So, viability is the sum of % germination and % dormant seeds in a seed lot—and the viability of our example seed lot would be 100%. Viability provides growers and seed companies with important seed-quality information, identifying the potential germination of a seed lot.

Germination

The purpose of a germination test is often confused with that of a seed vigor test. Seed germination is defined by the Association of Official Seed Analysts (AOSA) as "the emergence and development from the seed embryo of those essential structures, which . . . are indicative of the ability to produce a normal plant under favorable conditions." The purpose of a germination test is to determine how seeds perform under favorable conditions—conditions seldom encountered in the more stressful conditions of a greenhouse or field environment. These favorable

conditions are specified by AOSA in the Rules for Testing Seeds. Because seeds are always tested for germination under favorable conditions, it isn't surprising that growers seldom achieve germination percentages reported on the seed label in the greenhouse. This is seen in our example, where 95% of the seeds germinated under the favorable testing conditions in a laboratory, but only 80% emerged in the greenhouse.

Based on this, what is the practical value of a seed germination test? First, it's a legal requirement: The Federal Seed Act requires that a germination test be provided on most agricultural seed marketed across state boundaries. (Flowers are an exception.) Such a requirement allows purchasers to interpret the quality of the seed. It also assures that results on seed labels represent what's in the seed packet, a process known as "truth-in-labeling." Companies use favorable conditions because they're the most reproducible. Because germination is a legal requirement, any dispute concerning seed quality must be based on techniques that provide reproducible results.

Second, because a seed germination test is conducted under favorable and reproducible conditions, results from two different laboratories testing the same seed lot should be the same. This is important for marketing seed nationally and internationally, where seed is produced and tested in one place, but sold and tested in another. Seed buyers are assured that seed quality is the same as the seed company advertises under standardized, reproducible conditions. Seed germination minimizes concerns about seed quality and permits orderly seed shipments from production location to area of use.

Vigor

Seed vigor can be defined as the seed properties which determine the potential for rapid uniform emergence and development of normal seedlings under a wide range of field conditions. Because a seed vigor test is a more sensitive measure of seed quality, its objective is to more accurately portray how a seed lot will perform under the unpredictable environment of commercial greenhouse production. In our example, a vigor test should provide an 80% emergence result. Unlike germination testing, vigor testing isn't a legal requirement, so seed companies can design any vigor testing protocol that best mimics the actual planting environment that seed encounters during plug production. For example, this may be nine- and twenty-one-day counts of usable seedlings in a plug tray or computer imaging using the increase in size of seedling cotyledons during emergence in a standard plug tray. Another approach is an accelerated-aging test, which exposes seeds to high temperatures and high relative humidity stress for short time periods. Poorer quality seeds deteriorate faster under these conditions. Vigor test results are important to growers who depend on rapid, uniform emergence and have a goal of 100% usable seedlings. Seed vigor also provides important benefits to seed companies, helping them identify seed lots that are the most vigorous for further seed enhancement technologies. Improved inventory management and the identification of the highest quality seed lots for marketing are also advantages.

September 1996.

⚜

Genetic Testing for Seed Vigor and Uniformity

by Miller B. McDonald

As demand for flowers has increased, production techniques have changed. Initially, producers grew flowers in pots or rows in a tray, and flower seed performance wasn't that important. However, recent emphasis on plug production has created new, highly defined demands for seed performance.

Complete, rapid, and uniform emergence of seedlings is essential for successful flower plug production. While the ideal of 100% usable seedlings in a plug tray remains an elusive goal, less than 100% means unused greenhouse space and lost grower revenue. Rapid emergence is essential for faster greenhouse turnaround and to free up space for more plants. Uniform emergence allows for more accurate timing of shipments and creates a more desirable appearance to the buyer.

Flower seed companies are keenly aware of these important requisites and recognize that enhancements in seed quality are vital to achieving these objectives. Improvements in seed quality can be made at two fundamental levels: genetic and physiological. The genetic level is where breeders select for parental lines that produce hybrids with superior performance. The physiological level is where seed technologists improve production and storage environments and remove seeds that perform poorly in a plug before packaging. By far, the most important of these is genetic purity.

Genetic Purity

Any time a new variety is developed, the limited amount of seed available is increased to quantities sufficient to supply the needs of growers. As this seed is increased, breeders monitor the new plant generations to ensure that they possess the desired unique variety traits.

Genetic purity is the confirmation that the variety initially developed by the breeder is the same variety marketed to the grower. Genetic purity is so important that seed companies routinely test their seed using genetic purity tests to make sure it meets two objectives: assuring that the stated variety is present in the seed lot, and determining the percentage or number of seeds in the seed lot which indicate the genetic diversity inherent in the variety.

In self-pollinated crops where pure-line varieties are produced, the second objective is of equal or greater interest, since it reflects the level of outcrossing

or selfing and the degree of uniformity that can be expected from a particular seed lot. Since most flower crops are hybrids, it's important to assure the
success of hybridization.

Hybrid Seed Production

Hybrid seed production offers many important advantages to flower growers. From a consumer perspective, more vigorous plants are available, often
leading to larger, more colorful flowers. From a grower perspective, hybrid
seeds produce more rapid and uniform seedling emergence because all seeds
have the same genetic heritage. How does hybrid seed production lead to
uniform seedling emergence?

Hybrid seed production begins with developing parental lines that have
desirable genetic traits. Self-pollinating a line for a number of generations
creates inbred parental lines. After five to seven generations, the line becomes
genetically pure or homozygous (has identical pairs of alleles that are either
dominant or recessive such as AA or aa, respectively).

As this inbreeding continues, the lack of genetic diversity causes the
inbred parent to be more susceptible to environmental stresses. Thus, the
resulting inbred plants are often unattractive and unmarketable. However,
when the appropriate inbred parents are crossed (AA x aa, for example),
the progeny all possess the same genotype (Aa). This results in seedlings
and plants that are uniform in appearance and performance and very
adaptable to environmental stresses. These desirable responses are known
as hybrid vigor.

Not all agricultural crops display hybrid vigor. However, breeders have
successfully developed many hybrid flower crops that should result in seed
that has complete, rapid, and uniform emergence compared to seed
produced from open-pollinated varieties. These performance advantages do
come at a cost. Hybrid seed production is expensive. In most cases, the
process of crossing two inbred parents must be carefully monitored to ensure
that hybridization occurs. This requires emasculation of the female parent
and successful pollination by the male in a controlled environment. Both
processes are usually done by hand and are very labor intensive. In addition,
producing and maintaining inbred lines are also labor intensive. This is why
many hybrid flower seed crops are produced in Central America, Asia,
or Africa, where production conditions are carefully controlled and labor
is inexpensive.

The success of hybridization is central to uniform seedling emergence in plugs. Because all seeds in the hybrid have the same genetic composition, they respond to the environment similarly. To assure this genetic uniformity, most genetic purity tests have traditionally used morphological (physical) features of the plant for discrimination. However, the increasing proliferation of new flower hybrids and the limited number of morphological traits available to the breeder make accurate identification of hybrids difficult. As a result, seed companies have resorted to separations of seed proteins on electrophoretic gels. However, in many cases the small size of the seed provides little enzyme activity, and the small number of enzyme assays available still make genetic purity determinations difficult and challenging.

The ideal genetic purity test should detect the sequence of bases found on the DNA molecule that characterizes the hybrid. One technique to accomplish this is called Random Amplified Polymorphic DNA (RAPD). This approach takes small quantities of DNA from a single seed and increases it, using enzymes in combination with the appropriate building blocks of DNA, to quantities that can be resolved on an electrophoretic gel. The DNA is visualized on the gel as bands that represent the DNA in the seed. Hybrids with the same genetic composition have the same banding patterns. Hybrids with differing genetic composition have differing banding patterns.

Further, every seed in the hybrid seed lot should have the same banding pattern to have uniform performance. For example, in one test we examined ten individual seeds in a hybrid cyclamen seed lot and found that this wasn't the case. One possible explanation was that the inbred parents weren't bred to homozygosity and thus weren't genetically uniform. An examination of the RAPD bands produced from the ten individuals from the male line and ten individuals from the female line demonstrated that parental lines weren't genetically pure.

Remember that these findings represent only one seed lot from one flower crop. Still, we believe they have important implications for seed companies and growers that could extend to other floral crops. First, the inbred parents weren't bred to homozygosity. This means that crossing inbred parents produced a hybrid that was probably less vigorous than one that would have been achieved if the inbreds were genetically pure. Second, the genetic variability found in hybrid seed contributes to the lack of uniformity in seedling performance in plugs.

Until now, seed companies haven't had the ability to directly monitor the genetic composition of inbred lines. Using morphological features alone as the main criteria has led to development of genetic diversity within inbred lines and production of hybrid seeds of varying performance. RAPD markers have the potential to help breeders quickly improve the genetic purity of their inbred lines and increase emergence and uniformity of hybrid seed. It's now clear that continuing advances in seed technology focusing on seed enhancements will be limited until the genetic purity of hybrid flower seed is achieved. Using RAPD markers can help to identify and improve the genetic purity of flower seed crops.

Acknowledgments: Appreciation is extended to Jianhua Zhang, visiting scientist from China, for conducting this research and to Patty Sweeney, The Ohio State University research scientist, for stimulating discussions concerning its importance.

Winter 1996.

New Test Rates Seed Performance under Stress
by Miller B. McDonald

In plug production, complete, rapid, and uniform emergence of flower seeds is essential. Each unfilled impatiens cell, for example, means a $.03 to $.05 market loss for the grower. As a result, plug growers are challenging the flower seed industry to provide seed lots with the highest quality and the potential to produce 100% usable seedlings in a tray. The seed industry has responded by initiating seed-quality-control programs to test seed vigor in flower seed lots.

Seed Vigor Tests

Seed vigor can be defined as the seed properties that determine the potential for rapid uniform emergence and development of normal seedlings under a wide range of field conditions. A seed vigor test's purpose is to accurately portray how a seed lot will perform under the varied and less-than-optimal conditions of commercial greenhouse production. This contrasts with a germination test, which is conducted under optimum conditions to achieve reproducible test results, thereby ensuring statewide and nationwide marketability. How seed vigor and germination tests differ is depicted in Figure 1. Because germination tests are conducted under favorable conditions, weak seeds continue to germinate for longer periods. Higher germination values with increasing seed deterioration result and that's why

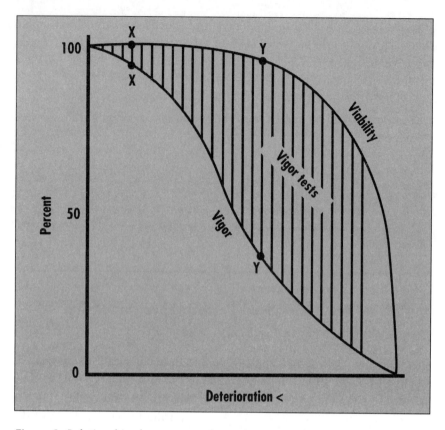

Figure 1. Relationships between seed germination and vigor with increasing seed deterioration. Note that X and Y represent two seed lots with approximately the same germination percentage but very different seed vigor due to deterioration.

few growers actually achieve a greenhouse emergence value equivalent to the germination result.

Most vigor tests, in contrast, utilize some form of stress (cold, drought, high relative humidity, etc.) during germination that makes weak seeds ungerminable. This results in vigor test values that correspond with the level of seed performance expected in the greenhouse.

Accelerated-Aging Test

The accelerated-aging test stands as perhaps the most universally adaptable seed vigor test. This test subjects seeds to high temperature (106°F) and high relative humidity (about 100%) stress for short durations (usually seventy-two hours), followed by germination under the Association of Official Seed

Analysts (AOSA) Rules for Testing Seeds recommended conditions. Under these conditions, low-quality seeds deteriorate more rapidly than high-quality seeds. The seed industry uses this information to accurately portray seed lot performance under stress conditions and to determine storage potential so that seed inventory can be better managed.

Preliminary studies with small-seeded flower crops such as impatiens using standard testing protocol demonstrated that flower seeds deteriorate too rapidly to achieve meaningful vigor test results. Researchers attribute this rapid deterioration to small seed ready uptake of water in the 100% relative humidity environment, causing premature seed death.

Replace Water with Salt

Because various saturated-salt solutions modify and balance relative humidity in a closed container (Table 1), one way to delay seed moisture uptake would be to substitute a saturated-salt solution for water in the accelerated-aging chamber. To test this, saturated-salt solutions of potassium chloride, sodium chloride, and sodium bromide were

Table 1. Percent relative humidity in various saturated-salt solutions at 68°, 77°, and 86°F.

Salt solutions	Temperature		
	68°F	77°F	86°F
	% Relative humidity		
Lithium chloride	11.2	11.2	11.2
Potassium acetate	23.2	22.7	22.0
Magnesium chloride	31.5	32.9	32.4
Potassium nitrite	49.0	48.2	47.2
Calcium nitrate	53.6	50.4	46.6
Sodium nitrate	65.3	64.3	63.3
Sodium chloride	75.5	75.5	75.6
Ammonium sulfate	80.6	80.3	80.0
Potassium nitrate	93.2	91.9	90.7

prepared at 106°F and produced relative humidity values in the accelerated-aging chamber of 87%, 76%, and 55%, respectively. Goldsmith Seeds, Gilroy, California, supplied two impatiens seed lots that had high germination (96% or higher) values but which differed in seed vigor based on percent germination on a thermogradient table (Table 2). Impatiens 1843 was higher in seed vigor than impatiens 1842. Researchers conducted accelerated-aging tests according to the procedures described in the *AOSA Vigor Testing Handbook* with two modifications.

First, because the wire-mesh screen was too large to support the impatiens seeds, a fine nylon fabric was placed over the wire-mesh screen. Nylon prevented the wicking of free water into the seeds during aging. Second, the

Table 2. Percentage germination of two impatiens seed lots at differing temperatures on a thermogradient table for three to ten days.

	Temperature (°F)									
	57.2	60.8	64.4	67.1	69.8	72.5	76.1	78.8	82.4	86
	% Germination									
Impatiens 1842										
Days										
3	0	0	0	0	8	28	60	36	24	0
5	0	0	24	52	84	82	92	74	66	2
7	0	10	60	82	92	88	94	76	74	2
9	2	52	86	82	94	92	96	78	76	6
10	22	72	86	82	96	92	96	80	78	6
Impatiens 1843										
3	0	0	0	12	40	60	66	64	20	2
5	0	4	22	78	90	94	88	86	56	10
7	0	38	80	92	94	96	100	92	58	10
9	18	70	94	98	96	96	100	92	58	12
10	30	86	98	98	96	96	100	92	58	12

three saturated-salt solutions were substituted for water and 60 ml of each added to the inner chamber. Aging at 106°F was imposed on each seed lot for forty-eight, seventy-two, and ninety-six hours. After these periods, the seeds were removed. Four replicates of fifty seeds each were germinated at 68°F and evaluated for germination at four, six, eight, and ten days. Researchers repeated the study three times.

The saturated salts did reduce seed moisture uptake during accelerated aging. Impatiens seed lots equilibrated at 10.6%, 9.2%, and 6.5% seed moisture content for the saturated potassium-chloride, sodium-chloride, and sodium-bromide solutions, respectively, after only three hours aging. This contrasts to seeds in a standard AA test (using water), which had seed moisture contents as high as 77% after forty-eight hours of aging. As expected, the low seed moisture contents delayed seed deterioration, which was greater with potassium chloride followed by sodium chloride and then sodium bromide (Table 3). The higher the AA environment's relative humidity, the greater the seed deterioration rate. In addition, the longer the AA regime, the greater the seed deterioration. It should be noted that all three saturated-salt solutions easily differentiated the seed vigor of impatiens 1842 and impatiens 1843.

Identifying an optimum salt solution and aging time that allowed the test to be conducted over a sufficient period to differentiate seed lot vigor was a central objective of this research. From a practical perspective, the test period shouldn't be shorter than twenty-four hours, because any problems in removing the samples from accelerated aging in a timely fashion would strongly impact the final results. A more appropriate period would be between seventy-two and ninety-six hours aging, as a difference of plus or minus one hour would have less effect on results.

Table 3. Percentage germination of two impatiens seed lots after 4, 6, 8, and 10 days after accelerated aging for 48, 72, and 96 hours in saturated potassium chloride (87% RH), sodium chloride(76% RH) and sodium bromide (55% RH) salt solutions at 41°C.

| | | IM 1842 | | | IM 1843 | | |
| | | 48 hr | 72 hr | 96 hr | 48 hr | 72 hr | 96 hr |
Salt	Days Germinated	% Germination					
Potassium chloride	4	22	8	0	62	43	20
	6	37	20	13	78	62	49
	8	45	25	16	86	66	58
	10	48	29	17	86	69	58
Sodium chloride	4	39	25	27	85	75	70
	6	64	60	45	92	88	83
	8	71	66	51	93	89	84
	10	74	69	53	94	89	85
Sodium bromide	4	43	49	48	86	80	83
	6	73	70	64	96	92	92
	8	82	76	75	98	96	94
	10	84	79	78	98	97	95

A final consideration: The sooner seed vigor differences can be detected following germination, the better. This allows seed testing laboratories faster turnaround of samples and provides vigor test information more quickly to the grower. Based on these criteria, the greatest differences in seed quality between the two seed lots were detected after four days germination following aging for forty-eight hours using sodium chloride or four days germination following aging for seventy-two hours using sodium chloride (Table 3).

Future Potential

These findings show that it's possible to manipulate relative humidity of the air during an accelerated-aging test. Small-seeded crops, such as impatiens, rapidly achieve equilibrium moisture content after only three hours in an accelerated-aging chamber using saturated-salt solutions. Reduced relative humidity of the surrounding air delays seed deterioration for as long as ninety-six hours, permitting greater uniformity of seed aging and flexibility in laboratory scheduling.

This new approach to accelerated aging has potential application for many other small-seeded flower crops that previously weren't tested for seed vigor using standard accelerated-aging protocols. Saturated potassium-chloride and sodium-chloride solutions should be used to determine the optimum aging period for each crop. Using the saturated-salt accelerated-aging test in flower seed quality testing programs will likely result in a seed product superior to that marketed today.

Appreciation is expressed to the Fred C. Gloeckner Foundation for partial funding of this research and Jianhua Zhang, visiting scientist from China, for technical support.

Winter 1996.

Store Your Seed the Right Way

by David Cross and Dr. Roger C. Styer

Seed quality is one of the most talked about subjects among plug growers everywhere, and storage is the key to maintaining high seed quality from seed harvest to sowing. All growers have seed in storage for various periods of time. Many order their seed for the whole season at one time, then use it throughout the season. Sometimes, seed is left over for next year. The effects of storage on seed quality depend on the storage conditions provided. For large plug growers, seed inventory can be a costly investment. Any seed-quality losses can translate into poorer seed performance in the plug.

Crops can store for different lengths of time. Table 1 lists some major crops by their relative storability. Use this table as a guide for how long seed can be reasonably stored under normal but not ideal conditions. When seeds have been enhanced or otherwise altered, their storage life will usually be shortened. This list includes pelleted, coated, primed, detailed, defuzzed, dewinged, or scarified seeds. Storing under more ideal conditions can extend storage of enhanced seeds.

Table 1. Relative Storage Life of Seeds*

Short	Medium	Long
Anemone	Ageratum	Amaranthus
Asparagus fern	Alyssum	Daisy, Shasta
Aster	Cauliflower	Stocks
Begonia	Celery	Sweet Pea
Browallia	Celosia	Tomato
Delphinium	Coleus	Zinnia
Herbs	Cyclamen	
Impatiens	Dahlia	
Lettuce	Dianthus	
New Guinea impatiens	Dusty miller	
Onion	Eggplant	
Pansy	Geranium	
Pepper	Lisianthus	
Phlox	Lobelia	
Salvia	Marigold	
Vinca	Petunia	
Viola	Portulaca	
	Snapdragon	
	Verbena	
	Watermelon	

*Adapted from *Principles and Practice of Seed Storage*, O.L. Justice and L.N. Bass, 1978, USDA Agric. Handbook #506.

What Makes up Seed Quality?

Viability is the ability of a live seed to germinate under optimum conditions. The ability of a live seed to germinate under a range of conditions and still produce a usable seedling is called vigor. These two terms aren't the same, but together they describe optimum seed quality. Plug growers today are demanding higher seed quality with emphasis on usable seedling. What they're really asking for is seed that not only has high viability but also high vigor.

The length of time seed stays vigorous is a function of genetics, storage conditions, and time. During storage, vigor decreases before viability, meaning total germination won't be the best indicator for how the seed is holding up in storage. Two lots of the same variety can have 90% germination in the lab, but one lot will already be declining on the vigor curve while the other lot retains high vigor. Once a seed starts moving down the vigor or viability curve, it can't be reversed, only slowed down. All seeds will eventually decline and die during storage. It's only a matter of time!

The storage conditions seed is exposed to determine how fast it moves along the vigor and viability curves. The two most important factors in seed storage are moisture content and temperature. Seed moisture content adjusts according to the relative humidity of the surrounding air. However, seeds differ in the way they adjust their moisture content to humidity. Seed composition, especially protein or lipid content, strongly affects its attraction to moisture. Of these two factors, seed moisture content is the most important.

Moisture Content and RH Count

In 1960 J.F. Harrington outlined the following rules of thumb for seed storage:

- Every 1% decrease in seed moisture content doubles the storage life.
- Every 10°F decrease in seed storage temperature doubles the storage life.
- The sum in degrees F and % relative humidity (RH) of good storage conditions should be less than 100.

Proper storage conditions maintain relative humidity levels between 20 and 40%, giving corresponding seed moisture contents of 5 to 8%, a range that's safe for many seeds. When seed moisture content drops too low (less than 5%), storage life and seed vigor may decline. When seed moisture content rises above 8%, aging or seed deterioration can increase. Deterioration involves increases in respiration, breakdown of storage reserves, decline in nuclear structure, reduction in cell membrane integrity, and other biochemical processes that all result in vigor and viability loss. Seed moisture contents above 12% promote growth of fungi and insects. Most seeds can't germinate until seed moisture contents rise above 25%.

Currently, most growers store and ship seed in hermetically sealed, plastic-laminated foil packets and bags or vacuum-packed cans. These containers provide excellent barriers against moisture moving in or out. Therefore, seed must be at the proper moisture content before being packed, and air in the containers should have an appropriately low RH.

Storage temperature should stay between 40° and 70°F. Pay special attention to the % RH, as humidity changes with air temperature. Most seeds can be stored below freezing if their moisture content is low enough (5 to 8%). However, you should only freeze seeds for long-term storage. Take care to acclimate frozen seed to room temperature without letting moisture condense on the seed. Store seeds such as delphinium, primula, cyclamen, pansy, and geranium at cool temperatures (42°F) for best storage life. Other seeds, such as impatiens, will store longer with lower temperature and RH.

At PanAmerican Seed, we compared storability of two seed lots of Super Elfin White and Super Elfin Red impatiens. All lots were the same general age. Storage conditions consisted of different relative humidities (thus different seed moisture contents) and temperatures. Samples were taken out of storage at intervals up to one year and sown into plug trays. Usable seedlings were counted after twenty-one days under normal greenhouse germination and growing conditions. Storing impatiens seed at 15% RH appears to be too dry and reduces usable seedlings. Storage temperature had a very strong influence on impatiens, especially with poor vigor seed lots. When both varieties were averaged together for an overall prediction of storage response, impatiens seed quality dramatically dropped under unfavorable temperature and RH levels. The best storage conditions for impatiens are around 42°F and 25 to 30% RH. Similar recommendations apply to pansy seed.

Plug growers can use a frost-free refrigerator or climate-controlled room set at 42°F to store all of their seed. Foil packets and metal cans will protect against moisture as long as they remain sealed. Once opened, you should prevent seeds from taking up moisture from the air. Relative humidity in a seeding area can easily exceed 80% with all of the moisture around from filled plug trays, drum seeders, and watering tunnels. Often, open seed packets are out all day in the seeding area while the seeder operator uses them to sow the day's production. Exposed seed can easily increase moisture content

2% in two hours under these conditions. If this process continues with the same seed lots over several weeks, seed vigor deterioration and reduction will occur, especially in sensitive crops such as impatiens.

Simply closing the foil packet and placing it in a refrigerator won't remove extra moisture seed has absorbed from the air. Growers should place seed in a water vapor-proof container that can be sealed (such as a Tupperware container or Mason jar) with a thin layer of silica gel desiccant in the bottom, and store it in a refrigerator. This desiccant layer should be no more than ¼ inch deep, as more desiccant causes % RH to be too low. When the desiccant's color changes from blue to pink, regenerate the silica gel by placing in an oven at 230° to 360°F until color changes back to blue. You can also use a microwave oven.

If a refrigerator is too small to store all of your seed, look into building a climate-controlled room or buying a walk-in cooler. This room should be very well insulated (R-value=30) and vapor-sealed, with a thermostat and humidistat for controlling conditions at 42°F and 25 to 30% RH. Make sure cooling unit fans are set on continuous operation to eliminate temperature fluctuations in the room. You can control humidity with a commercial desiccant dryer. Dehumidifiers for home use aren't suitable for controlling RH in a seed storage area because they can only reduce RH to 35%, and they generate heat.

Culture Notes, January 1996.

PGRs at Seeding Reduce Early Stretch

by Andrew Britten

One problem in growing some greenhouse plants from seed is early stretch—the rapid stem growth and elongation prior to the emergence of the plant's first leaves. This early stretch makes it more difficult to grow a full, compact plant—the kind of plant today's consumer wants to buy.

Take the marigold, for instance. It's a plant that stretches up early and can quickly become tall and lanky prior to the first plant growth regulator (PGR) application. Growth regulators generally are applied after the first true leaves have expanded. Recent trials, however, have shown that the growth regulator A-Rest when applied to the media after seeding will reduce stem height by half, resulting in a fuller, more compact plant. A-Rest was used because it's less potent and more forgiving than many of the growth regulators on the market today. The long-term effects on the plant are positive, and there's less chance of excessive stunting.

Discussions during Ball Publishing's International Plug Conference in October 1998 in Orlando, Florida, about the use of growth regulators at seeding prompted our interest in evaluating this technique here at Heartland Growers on some of the tough crops that characteristically give us problems with early stretch.

Shown here are snapdragons ready to transplant. Treated crops are on the right, and untreated crops are on the left. *Photo: Andrew Britten*

The Trials

In December 1998 we began our tests here in Westfield, Indiana, on plugs of several different plant varieties to test this concept of applying A-Rest to the media after seeding. We tried different varieties of marigold, snapdragon, zinnia, dahlia, portulaca, petunia, dianthus, celosia, and salvia. (See "A-Rest Results on Various Plants" on page 46.)

We used the A-Rest at three different application rates: the recommended foliar rate for each particular plant variety, double the recommended foliar rate, and half the recommended foliar rate. We treated four trays of each plant variety. Applications were made using a hydraulic sprayer. A normal spray application of A-Rest was administered across the top of trays, applied to the media and the seeds.

On the species we evaluated, we discovered that, overall, the best rate for A-Rest applied immediately after sowing the seed is 6 ppm. Rates varied depending on the species/cultivar, environmental factors, fertilizer regime, and other practices.

How It Works

The PGR doesn't need to be applied when the seed is sown. In general, it takes up to ten days before the seed coat cracks and roots begin to emerge. The PGR application is most effective prior to the cracking of the seed coat.

As soon as the roots emerge from the seeds, the A-Rest is absorbed. In effect, the plugs get "drenched" without the soil having to be saturated. This immediate effect from the PGR keeps the plant compact without adverse results on later development or growth. We've also found that the plugs treated with A-Rest rooted more quickly and more substantially than those untreated.

The Benefits

We've found that using growth regulators at the seeding stage on tough crops such as marigolds, snapdragons, and zinnia resulted in many benefits, including:

- Plants rooted more quickly and more substantially,
- Fuller, more compact plant trays and flats,
- Plant leaves were closer to the soil,
- Increased branching.

For a fourteen-acre operation such as Heartland Growers, which sells half of its plugs to growers and uses the other half for its own finished plants, producing a full root system that leads to a high-quality crop is very important not only to us, but to our wholesale customers.

Thanks to the success of our first trials, we plan to initiate additional research this summer on impatiens and seed geraniums.

September 1999.

A-Rest Results on Various Plants

by Andrew Britten

Marigold Safari Queen

Trays were sprayed with 6, 12, and 24 ppm of A-Rest. The 12 ppm rate produced the nicest looking finished plug. All applications decreased the length of the stem. The 12 and 24 ppm rates made the plant half the size of the untreated tray. The 24 ppm spray decreased the overall size of the leaves, leaving the plants looking initially stunted.

Snapdragon Floral Showers Scarlet

Trays were sprayed with 6, 12, and 24 ppm. The 6 ppm rate produced the nicest looking finished plug. Rooting was greatly increased on this crop.

Zinnia angustifolia Profusion Cherry

Trays were sprayed with 5, 10, and 20 ppm. The 10 ppm rate produced the best looking finished plug. The overall effects on zinnias were similar to

This is a comparison of zinnia at Stage 3 of treated (right) and untreated (left) plants.
Photo: Andrew Britten

those of marigold. The 20 ppm rate delayed blooming by one week.

Dahlia Figaro Mix

Trays were sprayed with 6, 12, and 24 ppm. The 6 ppm rate produced the nicest looking finished plug. All applications slowed down the stretch commonly associated with this crop, but the higher rates resulted in much smaller leaves. After transplanting, the leaves quickly expanded to normal size.

Portulaca Sundial Mix

Trays were sprayed with 5, 10, and 20 ppm. The 5 ppm rate produced the nicest looking finished plug. The higher rates kept leaves fairly small. Again, after transplanting the leaves soon expanded to a normal growth pattern.

Petunia Madness Burgundy

Trays were sprayed with 4, 8, and 16 ppm. We produced the nicest plugs at 8 ppm. This was the fullest and best rooted tray. The 4 ppm rate allowed too much stem elongation before the first leaves. The 16 ppm rate created small leaves.

Dianthus Telstar Crimson

Trays were sprayed with 6, 12, and 24 ppm. The 6 ppm rate was all that was needed to produce a nice-quality plug. The 12 and 24 ppm rates held the plug down harder than initially needed. Rooting was greatly increased in this crop.

Celosia Century Mix

Trays were sprayed with 3, 6, and 12 ppm. None of the rates decreased the length of the stem, but all rates did hold down the height of the plug after the first set of true leaves. However, we saw no hints that the A-Rest helped alleviate early budding.

Salvia farinacea Victoria Blue

Trays were sprayed with 3, 6, and 12 ppm. The 6 ppm rate produced the nicest finished plug. The 3 ppm rate allowed a small amount of stem elongation, while the 12 ppm rate kept the plants a little too tight. Even the highest rate quickly grew out of the application, but it didn't produce a very full-looking tray.

September 1999.

Progress Report: Bonzi on Seed

by Claudio Pasian and Mark Bennett

Some bedding plant crops, such as marigolds and tomatoes, tend to stretch very early after germination, especially if grown in low-light environments. By the time you apply growth regulators, the stretching of the hypocotyl has already occurred and sprays are ineffective against this early stretch.

At The Ohio State University, we speculated that an application of a plant growth regulator such as Bonzi to the seed before sowing could produce shorter plugs. We decided to test this hypothesis, aware that germination could be severely affected because the primary mode of action of paclobutazol, Bonzi's active ingredient, is inhibition of gibberellin biosynthesis—and gibberellins are needed for seed germination.

We decided to use Bonanza Gold marigold, Sun 6108 tomato, and Cherry Orbit geranium. We soaked the seeds for six, sixteen, and twenty-four hours in Bonzi solutions of 0, 500 and 1,000 ppm. After imbibition, we dried the seeds for twenty-four hours. After that, we handled and sowed the seeds as regular seeds. We measured seedling height and percent emergence sixteen, twenty-six, and thirty-six days after sowing.

To our surprise, germination was only partially affected in treatments at high concentration and long periods of soaking. In the case of tomato, we didn't notice any negative effect on germination, even at our highest plant growth regulator concentration. As you might expect, heights of seedlings decreased with increasing concentration of Bonzi. These preliminary results indicate that this method of growth regulator application (especially the six- or sixteen-hour soak in 500 ppm Bonzi) may be feasible and could benefit plug growers of marigold and other crops prone to stretching.

Do other crops respond the same way? We don't know but are in the process of conducting further tests. Our suspicion is that some seeds will be able to pick up enough growth regulator due to the characteristics of their seed coats. Others may have seed coats that don't allow any growth regulator to be attached or may have seed coats too permeable to the plant growth regulator, which could severely diminish germination. Only after our research is completed will we know for sure.

What are the potential benefits of this application method? First, stretching can be controlled early on, when spraying isn't possible. Second, this

These are tomato seedlings twenty-six days after sowing. The control is shown on the left; a seedling grown from seed soaked for sixteen hours in a 500 ppm Bonzi solution is in the center; and a seedling from seed soaked in a 1,000 ppm solution for sixteen hours is on the right.
Photo: Mark Bennett

method can be considered friendlier to the environment because the amounts of active ingredient you end up using are much lower than other methods, such as spraying or drenching. We haven't measured the amount of active ingredient that's carried by the seed, but it's quite possible that it's a very small amount.

Finally, a word of caution: This method of application isn't described in any of the plant growth regulator labels available today. Growers who use this method *do so at their own risk* because this method is still at an early developmental stage.

December 1999.

Fast Cropping Works!

by Will Healy

During the last few years, university researchers and growers have developed production methods that maximize the growth rates of bedding plants. The key, however, to fast cropping bedding plants isn't just growing plants quickly, it's also using the correct size plug for the time available. Using different plug sizes during the course of the season will optimize the date of flower and the number of turns per season you can achieve. Now, if only the weather would cooperate!

Success Factor No. 1—The Right Plug

The most critical factor for fast cropping annuals is starting with a mature plug that has been induced to flower. Fast cropping is based on the principle that growers only need to transplant the plug, unfold a few leaves, and expand the flower.

Flower initiation

In some species (i.e. impatiens) the flower buds must be visible, while in other species (i.e. petunia) the plant is reproductive but flower buds are not visible. To ensure a uniformly induced plug tray, precise management of the Stage 3 environment is critical. If plants are induced too early, the finished plants are very small, while delaying induction results in large plants with delayed or sporadic flowering. To promote uniform flower induction, growers are trying the following techniques.

Ageratum

Long days during Stage 3 will promote earlier flowering in ageratum even though plants won't appear reproductive at the time of transplant.

Begonia

Begonia flowers appear after a minimum number of leaves unfold. The plants don't appear to be photoperiodic.

Celosia

Flowering in celosia is induced by drought or excessive light. When seedlings are induced too early, it's difficult to produce a marketable finished flat.

Impatiens

The first three to four axillary buds on impatiens produce vegetative shoots, while all subsequent nodes have the potential to produce either flowers or vegetative shoots. Impatiens aren't photoperiodic, and flowering appears to be promoted by high light or slight water stress. Once flower buds are visible, avoid excessive water stress, which causes flower abortion.

French marigold

French marigold is a very difficult crop to precisely time, if flower induction isn't accurately controlled. They are classified as short-day plants. If plants don't receive a short-day treatment after March 15, flowering is delayed. When seedlings are exposed to short days at the cotyledon stage, plants are small and won't finish at an adequate size.

Pansy

Recent work in Minnesota shows that pansies are long-day plants. The response is similar to that observed with petunias.

Petunias

Flowering occurs once the petunia plants have initiated a total of eleven nodes, with four leaves unfolded. Providing long days prior to the four-leaf stage doesn't affect flowering, while delaying the long-day treatment delays flowering. Petunias are long-day plants when grown at temperatures above 68°F and day neutral if grown at temperatures below 68°F, so you must maintain temperatures above 68°F during the long-day treatment to promote flower initiation.

Salvia farinacea

Salvia farinacea has a very long juvenility and is difficult to initiate in the plug tray.

Salvia splendens

When *Salvia splendens* plugs are grown under long days and high light intensity, plants initiate flowers after four to six leaves are formed. Avoid premature flower initiation to prevent plants from finishing too small.

Plug size

Not all plug sizes are appropriate for fast cropping. Although some species grown in 800-count trays can be induced to flower, many growers have a problem filling out the finished tray because the leaf area of an 800 plug is too small. Most growers find 512 or 288 plugs the optimum size for fast cropping, depending on available growing time.

The time from sowing to flower is relatively constant throughout the year. When plants are grown in plugs and then transplanted, the total crop time is the same, although the crop time spent in the plug tray versus the finished

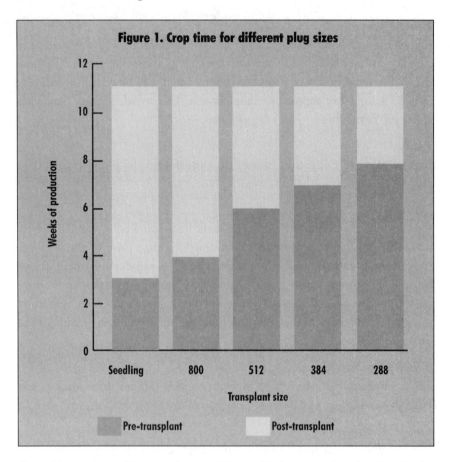

Figure 1. Crop time for different plug sizes

flat will vary depending on plug size (Figure 1). By managing the time in the plug tray versus finished flat, growers can accurately time the crop to hit specific sales windows.

To manage the time and space available to finish plugs, we developed a model that uses the Square Foot Week (SqFtWk) concept to manage time and bench space. Figures 2 and 3 show the effect of using different sizes of impatiens plugs on the date of flower and the number of turns in the crop time available.

Option A is the "cheap plug" strategy. By using 800 plugs followed by 512 plugs, only two turns are possible. The finish dates are mid-April and the end of May. It's difficult to hit earlier dates. With the long crop times used in this option, maintaining compact, high-quality plants can be difficult.

Option B, which uses larger plugs, provides more turns per greenhouse than Option A. The advantage of this combination is that the first ship date

Figure 2. Effect of impatiens plug size on flowering date and number of turns

Finish crop time (48 plant/tray)	800	512	384	288
Impatiens	7	5	4	3
SqFtWk per planting	800	512	384	288
Impatiens	27,998	19,999	15,999	11,999

Bench space available (sq. ft.)	4,000		
Number of flats	2,380	Length	Width
Flat size (sq. ft.)	1.68	11 in.	22 in.
Start date	March 1		
End date	June 1		
Production window (weeks)	13		
Square foot weeks (SqFtWk)	51,996		

is delayed, while the second and third ship dates hit during the peak shipping period.

Option C uses 288 plugs exclusively, allowing four turns from the greenhouse. If early markets are a possibility, then this option provides the greatest flexibility for finishing crops. Based on an early 1997 Easter, three crops of impatiens are possible if 288 plugs are used. The first planting is March 24, with the third crop shipped by May 30.

By matching the crop time for the different plug sizes with the time available a significant increase in gross sales is possible.

Success Factor No. 2—Cultural Program
Even if a fully mature and induced plug is used, the plant will fast crop only if your cultural program is intensively managed to maximize growth rate. Temperature and irrigation must be accurately managed to precisely time your bedding plant crop.

Temperature
There are three components of temperature: range, DIF, and average daily temperature (ADT). The temperature range is the maximum and minimum

Figure 3. Effect of impatiens plug size on flowering date and number of turns

Option A	Impatiens	Size	SqFtWk	Available SqFtWk	Plant	Ship	Total flats
	Turn 1	800	27,998	23,998	March 1	April 19	2,380
	Turn 2	512	19,999	3,999	April 19	May 24	4,760

Option B	Impatiens	Size	SqFtWk	Available SqFtWk	Plant	Ship	Total flats
	Turn 1	512	19,999	31,997	March 1	April 5	2,380
	Turn 2	384	15,999	15,998	April 5	May 3	4,760
	Turn 3	288	11,999	3,999	May 3	May 24	7,140

Option C	Impatiens	Size	SqFtWk	Available SqFtWk	Plant	Ship	Total flats
	Turn 1	288	11,999	39,997	March 1	March 22	2,380
	Turn 2	288	11,999	27,998	March 22	April 12	4,760
	Turn 3	288	11,999	15,999	April 12	May 3	7,140
	Turn 4	288	11,999	4,000	May 3	May 24	9,520

temperature at which plants will grow. Each species has a unique range. Vinca has the highest range while pansy has the lowest. When the temperature is above the maximum limit, crop quality declines. When the temperature is below the minimum, the plant stops growing.

DIF (the difference between day and night temperatures) is well understood by most growers. As the DIF increases more than about 8° to 10°F, plants stretch and growth regulators are required to maintain plant height. The more days with a DIF greater than 10°F, the more growth regulator required to control growth. When the DIF is consistently greater than 10°F, growers will use ARest, Sumagic, or Bonzi to control growth as these chemicals are more effective than B-Nine.

The ADT is the most critical temperature component when fast cropping bedding plants. Managing the ADT is a powerful tool growers can use to time the date of flower of most bedding plants. By increasing or decreasing the ADT, plant development is controlled to match the developing market periods. When the ADT is below 68°F, plant growth rate is reduced and development is delayed. A soil ADT of 68° to 70°F is critical when the plugs are transplanted. If the soil ADT isn't maintained, then root development is delayed and plants won't attain their growth potential.

Irrigation

During the course of the production program, you have a limited number of opportunities to irrigate or fertilize your crop. A vigorous root system is essential for fast cropping. Incorrect irrigation practices reduce root development, which limits your ability to push or hold the crop.

Just prior to transplanting, thoroughly water the plugs with a 75-100 ppm fertilizer solution of 20-10-20. This initial fertilization will soften up the plug and provide some nitrogen to promote growth once the plug is transplanted. If the plug is chlorotic use the higher rate. The ammonia in the 20-10-20 is important for promoting leaf expansion.

After transplanting, wet the soil thoroughly to promote strong root growth. If the plugs aren't sufficiently watered in, an interface between the plug and the potting soil will develop, restricting root development. If the soil doesn't have a nutrient charge, irrigate with 50-100 ppm of 20-10-20 fertilizer.

Once the plants have rooted out, continue fertilizing with 100-150 ppm of 14-0-14 (or 15-0-15) with every irrigation. The rate depends on the soil EC. Remember that soil EC is a measurement of what's left in the soil after the plant has extracted what it needs. If the EC climbs with time, reduce the ppm applied. If the EC drops, increase the ppm to maintain an EC of less than 1.0 (2:1 extraction).

Avoid using fertilizers containing more than 10% ammonium nitrogen as this will promote stretching of the plants. The initial fertilization with 20-10-20 plus the soil charge should provide enough ammonia to promote leaf expansion and branching. If the foliage color is pale or the leaf canopy doesn't close within ten days of planting, apply additional 20-10-20. Failure to apply sufficient nitrogen to the plants will result in poor root development, weak plants, and poor postharvest quality. Excessive nitrogen application will cause high soluble salts.

Growers have successfully used the fast crop program this past season to finish impatiens in ten days and petunias in fourteen days. When the right plug is grown under the optimum cultural program, several turns of high-quality plants are possible.

January 1997.

Improve Your Germination!

by Dr. Roger C. Styer

To solve germination problems with the majority of seed lots or varieties of a particular crop, plug growers need to concentrate on providing proper conditions for germination and early seedling growth. You can effectively group crops by temperature, moisture, or light requirements by following these guidelines.

Temperature

Table 1 lists plug soil temperatures for Stage 1 for some crops. If you're germinating crops during winter, you may need to use a plug chamber for 77°F while germinating the rest on the bench at 72°F. In the summer/fall season, germinate crops at 72°F or lower in a plug chamber, and germinate the rest on the bench as long as you can maintain reasonable temperatures. If you have more than one plug chamber, divide crops according to Table 1.

Table 1 - Plug Soil Temperatures for Stage 1	
77°F	Ageratum, Alyssum, Begonia (both), Browallia, Celosia, Coleus, Gomphrena, Hypoestes, Impatiens (less than 78°F), Lobelia, Nicotiana, Petunia (less than 78°F), Portulaca, Salvia, Verbena, Vinca.
72°F	Aster, Carnation, Cosmos, Dahlia, Dusty Miller, Dianthus, Gazania, Geranium, Lisianthus, Marigold, Pansy, Snapdragon, Stock, Viola, Zinnia.
67°F	Pansy, Phlox, Primula, Ranunculus (60°F), Viola.

Moisture

Moisture levels during Stages 1 and 2 are equally important to temperature. Table 2 shows what moisture levels many crops need. Wet means moisture is readily apparent to the touch and visible on surface as gleaming. Medium means moisture isn't visible on surface but is apparent to the touch, and the media isn't lighter colored. Dry means moisture isn't apparent to the touch, and media is lighter colored, with moisture contained mostly in the covering.

Light

When germinating in a chamber, use fluorescent lights for eighteen to twenty-four hours a day. Don't stack trays directly on top of each other, or

Table 2 - Plug Moisture Levels for Stages 1 and 2

Crop	Stage 1			Stage 2		
	Wet	Medium	Dry	Wet	Medium	Dry
Ageratum		X			X	
Alyssum		X			X	
Aster			X			X
Begonia	X			X		
Browallia		X			X	
Carnation		X			X	
Celosia		X			X	
Coleus		X			X	
Cosmos			X			X
Dahlia			X			X
Dianthus		X			X	
Dusty miller		X			X	
Gazania			X			X
Geranium	X				X	
Gomphrena	X					X
Hypoestes	X				X	
Impatiens	X				X	
Lisianthus		X			X	
Lobelia		X			X	
Marigold		X				X
Nicotiana		X			X	
Pansy	X				X	
Petunia		X			X	
Phlox			X			X
Portulaca		X				X
Primula		X			X	
Ranunculus		X		X		
Salvia	X				X	
Snapdragon		X				X
Stock		X			X	
Verbena			X			X
Vinca	X				X	
Viola	X				X	
Zinnia			X			X

light won't penetrate to the middle of each tray. Some growers wrap trays or carts of crops that need dark in black plastic and place them in a germination chamber at the proper temperature.

Major Crops

Following are key techniques for solving germination and early seedling growth problems for some major and minor crops.

Begonia

Keep begonia warm (75° to 80°F) and very moist also provide some light in Stage 1. Avoid plug mixes with high starter charges. Start early feeding (after cotyledon emergence) in Stage 2, and avoid drying out media surface until the first true leaf is ½ expanded. Roots will then go down through the cell. Provide some ammonium-nitrogen fertilizer after Stage 2.

Impatiens

Germinate impatiens at 73° to 75°F with light. Keep uniformly moist until Stage 2, then let surface dry off to get roots into media. Keep media EC low and pH high (6.0 to 6.2). Feed with more basic fertilizer after cotyledon emergence (13-2-13-6-3 or 14-0-14), and water early in the day to avoid tip abortion.

Petunia

For petunia, a temperature of 75° to 77°F is best with light. Keep medium moisture until Stage 2, then dry off the surface for best rooting. Keep pelleted seed moist in plug trays, but don't worry about washing away the pellet. Keep media pH less than 6.5. Feed early and often.

Vinca

During Stage 1, keep vinca warm, moist, and in the dark for best germination. Reduce moisture in Stage 2 to get root into media. Cover with coarse vermiculite if needed. Keep warm and dry; use low ammonium-nitrogen fertilizer; and keep media pH less than 6.5 and EC low for best seedling growth.

Verbena

Germinate verbena at 72° to 77°F on the dry side from sowing to end of Stage 2. Cover with coarse vermiculite to help keep humidity and air around the seed. Use ½ the amount of water with this crop as compared to other crops from beginning of germination to end of Stage 2.

Salvia

For salvia, keep a temperature of 75° to 77°F with cover, if needed. Keep moist during Stage 1, but reduce moisture in Stage 2 for best root growth. Use low EC media and less feed in Stage 2 and early Stage 3 than with other crops. *S. farinacea* takes longer than *S. splendens* to grow and should also be grown warmer.

Pansy

Pansy can germinate from 65° to 75°F, but must be moist during Stage 1. Cover if needed. Reduce moisture in Stage 2 for best rooting. Use primed seed for fall sowings. Keep media pH less than 6.5 and EC low. Watch early seedling stretch in the fall.

Geranium

For geranium, maintain a temperature of 70° to 75°F, no higher. Cover with coarse vermiculite and keep medium moisture. Reduce moisture in Stage 2 for best rooting. Keep media pH 6.0 to 6.2. Use HID lights after Stage 2 for best growth.

Snapdragon

Germinate snapdragon at 70° to 75°F with medium moisture. Reduce moisture levels in Stage 2. Use media with low EC and keep pH less than 6.5.

Minor Crops

Dusty miller

Most growers double-seed dusty miller. Keep temperature 70° to 75°F, and provide medium moisture, not saturated. Keep EC low, grow warm (65° to 68°F), and avoid overwatering to prevent disease.

Lisianthus

Germinate lisianthus at 72° to 75°F with medium moisture and light. Allow some drying of soil surface in Stage 2 as roots go deep very early (will improve algae control). Keep media pH 6.0 to 6.2, start feeding early, and grow under HID lights in winter. Best if grown with begonias in winter (for temperature and EC levels).

Portulaca

Keep portulaca warm (78° to 80°F) with medium moisture to germinate. Dry down soil surface in Stage 2 to prevent stretch and disease. Grow warm and use low ammonium-nitrogen feed and HID lights in winter.

Dianthus

Germinate dianthus at 70° to 75°F with medium moisture. Cover if needed. Use media with low EC and keep pH less than 6.5. Allow some drying of soil surface in Stage 2 for best rooting.

Celosia

Keep celosia warm to germinate (75° to 77°F) with medium moisture and low EC. Reduce moisture levels in Stage 2 to control early stretch. Avoid stress on plants to prevent premature flowering.

Phlox

Germinate phlox cool, dry, under a covering, and in the dark. Keep media EC low, but avoid stress on plants to avoid diseases.

Aster, dahlia, and zinnia

Germinate aster, dahlia, and zinnia cool, dry, and with a covering of coarse vermiculite. Avoid overwatering for disease control. Don't hold plugs of zinnia!

Culture Notes, February 1996.

The Ideal Germination Chamber
by Dr. Roger C. Styer

You can germinate your plug trays on your greenhouse benches, but a germination chamber can increase your germination rate by 10% or more compared with greenhouse benches when both systems are optimized. The advantages of using germination chambers include higher germination percentage and faster and more uniform germination. You'll use less greenhouse space and also have to pay less attention to your plugs to provide proper moisture and temperature levels during Stage 1. Disadvantages include the cost of building a good chamber, handling plug trays in the product flow, and timing the movement of plug from chamber to greenhouse to give the best germination without stretching your seedlings.

Why a Germination Chamber?

Germination chambers are environmentally controlled rooms in which seeded plug trays are placed to germinate. They are designed so that trays may be stacked vertically on movable carts or on racks and rolled or carried out of the room and into the greenhouse. Temperature is closely controlled with heating and cooling systems. Moisture is initially provided by watering in the plug trays directly after seeding and, once in the germination chamber, is maintained by using a fog system. Light may or may not be used. Oxygen

levels to the seed are improved by not oversaturating the seed with moisture or burying the seed in the plug medium. Environmental conditions can be monitored and maintained with simple thermostats and timers or with computers.

Designing the Chamber

There's no single perfect design for a germination chamber! Most growers build them out of exterior plywood and two-by-fours and improve them over time. Some buy a prefabricated cooler and turn it into a germination chamber. The type of chamber you build will depend on: 1) how much space you need to germinate plug trays, 2) how many different temperatures you need, 3) how long during the year you need the chamber, and 4) how much money you want to spend. I've included diagrams of a chamber design you can use as a guide for building or improving your own chamber. This design was set up with lights, but you can leave them out.

There are several key considerations when building a germination chamber. First, you need to calculate the capacity and size of the chamber based on how many plug trays you need to germinate in a given period of time and how many individual chambers you need. The types of crops in the chamber will also affect its size. For example, begonias take about a week to germinate and will take up chamber space longer than marigolds, which take only about two days to germinate.

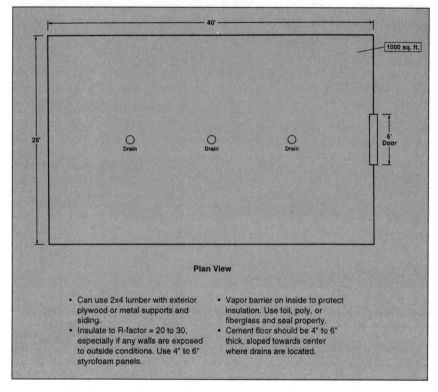

Plan View

- Can use 2x4 lumber with exterior plywood or metal supports and siding.
- Insulate to R-factor = 20 to 30, especially if any walls are exposed to outside conditions. Use 4" to 6" styrofoam panels.
- Vapor barrier on inside to protect insulation. Use foil, poly, or fiberglass and seal properly.
- Cement floor should be 4" to 6" thick, sloped towards center where drains are located.

Second, make sure the chamber will be at least eight feet high if no cooling system is needed and at least ten feet high if a cooling system will be hung from the ceiling. This height is needed for airspace above the carts or racks that allows the fog to flow without condensing on the plug trays. Too often growers build their chambers too low, with the result that the plug trays on top get too wet from the fog.

Third, plan to slope the chamber ceiling to prevent condensation from dripping onto the plug trays. Condensation will form on the ceiling and needs to be channeled away to the sides before water drops form.

Fourth, insulate the germination chamber to an R-value of at least twenty to help maintain temperature inside regardless of where the chamber is located. Various materials can be used for insulation, with the R-value depending on the type and thickness of material used. Don't forget to insulate the floor.

Finally, the inside of the chamber needs a moisture barrier. This material will protect the insulation from getting wet, which reduces its R-value. Also, moisture will stay inside the chamber where it belongs. Examples

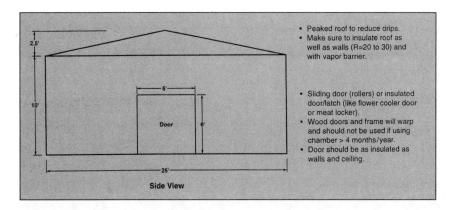

- Peaked roof to reduce drips.
- Make sure to insulate roof as well as walls (R=20 to 30) and with vapor barrier.

- Sliding door (rollers) or insulated door/latch (like flower cooler door or meat locker).
- Wood doors and frame will warp and should not be used if using chamber > 4 months/year.
- Door should be as insulated as walls and ceiling.

2.5'

10'

6'

Door

6'

25'

Side View

of moisture-resistant materials include polyethylene film, aluminum foil, and fiberglass.

Controlling the Environment

The best method of accurately heating a germination chamber is to use hot water pipes around the baseboards, connected to a separate hot water heater. You need to calculate how much heating capacity this system needs for the size of chamber, number of chambers, insulation provided, size of pipe used, outside temperature, and maximum inside temperature desired. If you're lighting your chamber, be sure to account for the extra heat from the lights and ballast.

For cooling the chamber, use cooling units designed for high humidity and low airflow. This type of unit is commonly used in meat packaging rooms, vegetable and fruit storage rooms, and flower storage rooms where high humidity is needed. Unit coolers can be mounted flush to the ceiling, pulling warm air up and discharging cool air through coils on each side of the unit. Allow about 1½ feet for these units to hang from the ceiling. Remember, these cooling units will set up their own airflow patterns, so your fog system will need to be adjusted accordingly. Contact your local heating/cooling contractor for more information.

Moisture control is critical for germination. Plug trays should not be hand-watered in the chamber because seed could be buried or washed out of the cells. The proper amount of water should be added to the tray after seeding but before going into the chamber. A fog system (ten-micron droplet size) will maintain high humidity inside the chamber to keep the original moisture level in the plug tray without adding or reducing moisture. The finer and more uniform the droplet size, the better the fog distribution and

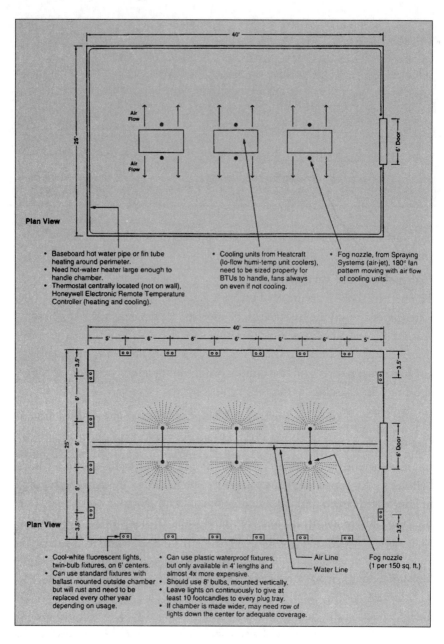

- Baseboard hot water pipe or fin tube heating around perimeter.
- Need hot-water heater large enough to handle chamber.
- Thermostat centrally located (not on wall), Honeywell Electronic Remote Temperature Controller (heating and cooling).

- Cooling units from Heatcraft (lo-flow humi-temp unit coolers), need to be sized properly for BTUs to handle, fans always on even if not cooling.

- Fog nozzle, from Spraying Systems (air-jet), 180° fan pattern moving with air flow of cooling units.

- Cool-white fluorescent lights, twin-bulb fixtures, on 6' centers.
- Can use standard fixtures with ballast mounted outside chamber but will rust and need to be replaced every other year depending on usage.

- Can use plastic waterproof fixtures, but only available in 4' lengths and almost 4x more expensive.
- Should use 8' bulbs, mounted vertically.
- Leave lights on continuously to give at least 10 footcandles to every plug tray.
- If chamber is made wider, may need row of lights down the center for adequate coverage.

Air Line
Water Line

Fog nozzle
(1 per 150 sq. ft.)

the better the temperature uniformity within the chamber. Too much air movement can cause uneven fog distribution, resulting in wet and dry spots.

The best fog system for a germination chamber uses AirJet Fogger Nozzles, from Spraying Systems Co.; an air compressor; air and water lines

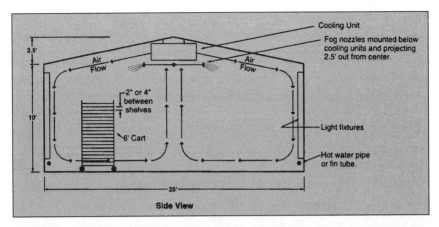

Cooling Unit

Fog nozzles mounted below cooling units and projecting 2.5' out from center.

2.5'

Air Flow

Air Flow

2" or 4" between shelves

10'

6' Cart

Light fixtures

Hot water pipe or fin tube.

25'

Side View

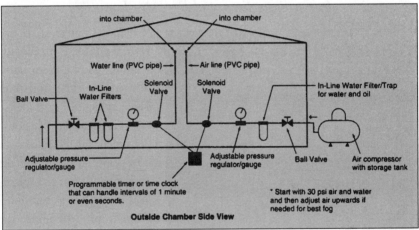

into chamber

into chamber

Water line (PVC pipe)

Air line (PVC pipe)

In-Line Water Filters

Solenoid Valve

Solenoid Valve

In-Line Water Filter/Trap for water and oil

Ball Valve

Adjustable pressure regulator/gauge

Adjustable pressure regulator/gauge

Ball Valve

Air compressor with storage tank

Programmable timer or time clock that can handle intervals of 1 minute or even seconds.

* Start with 30 psi air and water and then adjust air upwards if needed for best fog

Outside Chamber Side View

at 30 psi; solenoids; and a timer (see diagram). These air-over-water nozzles have a larger orifice that keeps them from clogging up. Output from these nozzles is much higher than from high-pressure nozzles, which means fewer nozzles per chamber (one nozzle/150 ft²). Use a timer or humidity sensors that can operate accurately above 85% relative humidity to control the amount of fog.

You don't have to light your germination chamber, but it helps. If no lights are used, then crops need to be monitored more closely for root emergence and then moved into the greenhouse on time for growing on. If left too long in the chamber the seedlings will stretch rapidly, and the crop will be ruined. If taken out too soon before germination has really started, total germination and uniformity will be reduced. Establish a set number of days for each crop, and sometimes for certain varieties, for when the trays should be pulled out of the chamber. Check them twice a day to make sure you don't miss anything.

Remember, a good germination chamber meets a few simple requirements: It should control temperature and moisture accurately for all crops and tray sizes and be large enough for your needs.

Winter 1996.

Successful Transplanting— Surviving the Critical First Week!

by Dr. Roger Styer and Dr. Dave Koranski

Whether you're growing your own plugs or buying them in, getting them transplanted on time and with good results is not always easy. Good-quality plugs can be ruined during holding or transplanting, or during the critical week after transplanting. If plugs are not toned properly or held under the proper conditions, they will quickly stretch, lose their roots and root hairs, become diseased, and, generally, will not be of good quality for transplanting. Poor-quality plugs, on the other hand, rarely make a good-quality transplant—no matter what the grower does!

You need to pay attention to many of the same factors for successfully transplanting plugs as for growing plugs. These include: water quality, media quality, environment, nutrition, moisture management, height controls, and diseases. Even simple factors, such as how deep to plant the plug, pulling or dislodging them from the plug trays, feeding them immediately after transplanting, and at what temperature to grow them can make the difference between rapid and uniform takeoff after transplanting or uneven growth and flowering.

Transplanting

The basics of water and media quality should be addressed before transplanting starts. Water quality for bedding plant crops should be: alkalinity less than 120 ppm, EC less than 1.0 mmhos, sodium less than 40 ppm, and chlorides less than 80 ppm. The soilless mix should be consistent and repeatable and have the following attributes: pH 5.5 to 5.8 (except for geranium, lisianthus, marigold, and impatiens, which need pH 6.0 to 6.2), soluble salts less than 1.0 mmhos (2:1 dilution), good water-holding capacity, and air porosity of 15 to 20%. If using some soil in the mix, then the starting pH can be 0.5 units higher. A complete soil analysis should be made prior to planting, with pH and EC monitored regularly during the crop cycle.

Whether transplanting is done manually or mechanically, predibble the containers to reduce the potential of root damage to plugs. Media should be somewhat moist to retain the shape of the hole. Make sure to adjust the dibble for the plug size you are transplanting, no deeper! Use a plug dislodger or popper to prevent breakage or tearing the plug by pulling. Choose the correct templates for plug dislodgers for each plug tray size that you use. Plugs should be watered two to three hours prior to transplanting to make extraction from the tray easier. Place the plug in the soil only slightly deeper than the top of the root ball, and firm the soil around the plug. If the dibble hole was made correctly for the plug size used, you may not need to firm the soil around the plug. The first watering will settle the media around the plug's roots. Separating multiple seedlings in a plug will damage seedling roots and isn't recommended! Water flats with clear, good-quality water immediately. No fertilizer or fungicide should be applied until after the roots have started to root out in approximately three days.

Whether you transplant by hand or by machine there is no excuse for transplanting poor-quality plugs! Train your crew to watch for and remove plugs that are poorly rooted, diseased, or too stretched. Avoid transplanting plugs of various sizes from the same plug tray into the same bedding flat. Quality control starts before you put the plug into the bedding flat.

First Week after Transplanting

Temperature

Initially, transplants should be allowed to root at 65° to 70°F soil temperature for the first three to five days. Once established, soil temperature can be lowered to 60° to 65°F nights. Growers should use caution at temperatures of 50° to 55°F, as plant stunting and flower delay can occur on plants with warm temperature requirements. When measuring and controlling temperature, remember that the difference between air and soil temperature can be as much as 5° to 10°F, depending on the light levels, plant canopy cover, and whether growing on unheated floors.

Water

Moisture can be applied manually or via automatic systems. The amount of water must be uniform enough to thoroughly saturate the growing media and allow leaching of 10 to 15%, particularly when dealing with less than desirable water quality. The growing media should then be allowed to dry down between waterings for good root growth. Remember, water quality is important, so have your water tested at least every six months.

Nutrition

If your growing media has a starter fertilizer charge in it, avoid feeding with water-soluble fertilizers until roots have become established (after five days). The type of fertilizer, rates, and frequency to use depends on the crop, cultivar, growing conditions, growing media, and the type of finished crop desired. Feed levels generally range from 100 to 300 ppm nitrogen. When using a slow-release fertilizer in the growing media, supplemental feeding may or may not be needed, depending on how much water goes through the containers and the growing temperature. Avoid using slow-release fertilizers in northern spring seasons, because you can lose control of your plant growth once the weather warms up.

Use low ammonium-nitrogen fertilizers if the soil temperature is less than 65°F average. Soil bacteria are needed to convert the ammonium nitrogen/urea to nitrate nitrogen that the plant can use. These bacteria aren't as active when the soil temperature decreases below 65°F, which reduces the conversion process. Ammonia toxicity may result, causing various forms of plant damage (death of roots, leaf symptoms, more root rots, stunted growth).

You can promote good root growth while controlling shoot growth by using fertilizers with low ammonium nitrogen/high nitrate nitrogen, high calcium and magnesium, and low phosphorus (13-2-13-6-3, 14-0-14). Monitor your media pH, because these fertilizers tend to raise it. If you need to push the crop, then use more ammonium-nitrogen fertilizer, such as 20-10-20. Monitor EC levels; some crops are considered light feeders (vinca, impatiens, pansy), and other crops are known as heavy feeders (petunia, gerbera, geranium). Soluble salts can build up three to four times immediately around the roots when the soil dries down, resulting in burning of roots and lower leaf symptoms.

Light

Constant fertilization at lower concentrations (75 to 100 ppm nitrogen) may be necessary to maintain sufficient plant growth under high-light conditions (more than 3,000 foot-candles). Generally, these plants will need more ammonium nitrogen. Fertilizing under different light conditions isn't necessarily dependent on the frequency of watering. More nutrients are required under higher light conditions to support growth. Under low-light conditions, slow plant growth will require fertilizing less frequently and with more nitrate nitrogen and calcium.

Fungicides and Chemical Growth Regulators

Managing moisture, temperature, nutrition, air movement, and pH and EC levels can prevent most diseases. However, fungicides may be necessary to control specific disease problems on specific crops. Apply any fungicide drenches after the roots have started to come out (about three to five days). Sanitation is a key element in disease control. Keep floors clean, weeds pulled, puddles of water under benches drained off, and algae controlled.

Manipulating water, nutrition, temperature, and light can control plant growth. When additional control is needed, chemical growth regulators may be necessary. Wait until the roots are out to the side of the containers before applying (about one week after transplanting). If plants are stretching too quickly in the first seven to ten days, you are probably overfeeding them or keeping them too moist.

Specific Situations

Non-stop begonias

Keep plants under long days (short nights) until April 1 to promote uniform growth and enhance flowering. Use daylength extension with HID lights or night interruption with incandescent lights for four to five hours (similar to mum lighting). If plants are not lit, they will form corms and go dormant, resulting in uneven growth and flowering.

Vinca

This crop needs to be grown warm (more than 68°F) and dry, at a low medium pH (5.5 to 5.8), at a low EC level, and fed with more nitrate-nitrogen fertilizer. Allow plants to get established before applying fertilizers, then apply at low rates. Vinca is not tolerant of high EC, and high fertilizer levels will damage roots, stunting growth when plants are first transplanted. Vinca does best when fed 100 ppm nitrogen from a fertilizer high in nitrate nitrogen. Be cautious of the effect that ammonium nitrogen can have on vinca roots!

Thielaviopsis

Thielaviopsis (black root rot) mainly infects pansy, petunia, and vinca when they are grown under stressful conditions. Symptoms include yellow bottom leaves and sometimes yellow, immature leaves. There is a loss of root hairs within two to four days, and roots take on a patchy brown to black discoloration. It can be confused with pythium, but there is no odor or loss of root sheaths at tips. Black chlamydospores may be seen on infected roots with a hand lens. The infected plant grows slowly, if at all, and may die.

To control thielaviopsis, reduce plant stress, practice good sanitation, and don't overdo water and fertilizer. Keep media pH 5.5 to 5.8 for these crops because this disease grows better at higher pH levels but the plants don't. Minimize temperature stresses on the plants through proper heating, cooling, and shading. For vinca, keep temperatures 68°F or warmer. For pansy, in the summer/fall, keep temperatures <80°F. Keep EC levels less than 1.0 mmhos (2:1 dilution) by not overfeeding and by clear-watering between feeds. Allow plants to dry down thoroughly between waterings for good root growth. Use recommended fungicides such as Cleary's 3336, Domain, Banrot, or Terraguard as drenches, and apply after roots have started coming out of the root ball (three to five days). Use clean containers and avoid contaminating media with dust.

Boron deficiency

Symptoms of boron deficiency on pansy and petunia include: stunting; newest leaves that are strapped, hardened, distorted, or mottled; terminal bud abortion; and proliferation of side shoots. Boron deficiency can be caused by boron leaching out of the media in hot, sunny weather when you must water plants more often. Usually, fertilization is reduced to prevent plant stretch in this type of weather. Boron can also be tied up at media pH more than 6.5 and when calcium levels are high.

To control or prevent boron deficiency, test your media and water regularly. Keep soil pH less than 6.5 for both petunia and pansy. Monitor calcium and boron levels in growing media, water, and plants. Boron needs to be 0.5 to 0.7 ppm in the media and 40 to 60 ppm in the tissue. During periods of warm weather and frequent watering, use Solubor at ¼ oz./100 gals. or Borax at ½ oz./100 gals. as a drench about one to two weeks after transplant to provide supplemental boron. Repeat in two to three weeks if needed.

Why aren't my plugs rooting?

If plugs aren't rooting out after five days there could be several reasons. High soluble-salt levels in the media from overfertilization or too much starter charge could be one cause. Test the EC levels and keep less than 1.0 mmhos (2:1 dilution) for the first week. Don't fertilize transplants until after roots start to come out.

Another cause could be too much moisture. This could be due to low air porosity in the mix or from overwatering. Reduce the frequency of watering to allow the media to dry down between waterings. Remember, the plug's roots are in the top half of the container for the first week and need some air to grow.

Poor water quality could be another problem. Check water for high alkalinity, high EC, and high sodium or chlorides. Reduce alkalinity to 120 ppm with acid injection if needed, and keep EC less than 1.0 mmhos, sodium less than 40 ppm, and chlorides less than 80 ppm for best root growth. Look into a different water source, or clean up your water if it contains high soluble salts and/or sodium. Low soil temperatures right after transplanting inhibit roots from coming out. Keep soil temperature 65° to 70°F for the first three to five days, then lower them to your normal growing temperature. Remember, vinca needs to be grown warmer throughout the crop cycle.

Another reason the plugs are not rooting could be that the plug root system was not healthy before transplanting. If the root system is damaged or diseased, it is likely to get worse after transplanting.

Why do my plugs already have buds or flowers?

Premature budding or flowering can be a major problem in some crops in getting uniform, vegetative growth after transplanting. Crops that have a central stem and single flower (celosia, French marigold, petunia, pansy) are most likely to have difficulties growing out after transplanting and will put all of their energy into making sure that that flower comes out. The result is no energy left for side branches to develop until after that first flower is finished. Crops that have multiple branches, flowers, or plants in the plug will have no problems growing out (impatiens, alyssum, begonia).

To get plugs to grow out after transplanting when they have buds or blooms already showing, you need to increase the feeding with ammonium-nitrogen fertilizers. Double the rate you would normally use and feed with more 20-10-20 until vegetative growth resumes. Reduce light levels (less than 2,500 foot-candles), increase the temperature, and keep media more moist (not as much drying out in between waterings) for the first two weeks. Readjust the environmental and cultural conditions to what you'd normally have for finishing bedding plants after the first two weeks. Be careful of promoting root rots with these treatments! Apply a fungicide drench of Banrot or Cleary's 3336/Subdue for best protection.

March 1996.

Damping-off Basics

by Dr. Mary Hausbeck

Pythium and rhizoctonia are fungi commonly associated with damping-off. Damping-off fungi can occur early in production by causing seed rot and attacking seedlings before they emerge from the soil. While you may think the reduced germination is due to poor seed quality, damping-off may be to blame. Post-emergence damping-off is more readily recognized because the fungus attacks at the soil line after the seedling emerges from the soil, causing water soaking and constriction on the lower stem. Seedlings then collapse at the constriction point.

Damping-off fungi can also infect cuttings, resulting in water-soaking stem discoloration and rotting, thus preventing root system formation. Rhizoctonia fungi typically cause a dull brown to dark brown rot on cuttings. When pythium is involved, the blight often has a shiny, coal-black appearance. On stock plants where botrytis is a problem such as geraniums and New Guinea impatiens, botrytis spores carry over on cuttings and cause a blight similar to that caused by rhizoctonia. However, fuzzy gray spore masses give botrytis away.

Sanitation is the keystone of preventing and managing damping-off. Healthy seedlings grown near plants that have damped-off may become infected and suddenly collapse after transplanting. Growing media carrying the damping-off fungus can transfer the disease from the infested flat to clean growing media.

When conditions are unfavorable for infection, damping-off fungi form a survival structure that may be found where used soil or plant debris accumulate, such as in used pots or flats, flat fillers, bench tops, and potting areas. Without proper disinfesting you can spread infested soil particles to clean areas of the greenhouse. Also, infested soil particles from greenhouse walkways and floors may be unknowingly transported throughout production areas by employees.

Fungicide drenches are an important tool in preventing damping-off and halting its spread. Fungicides such as etridiazole (Terrazole, Truban), metalaxyl (Subdue), and propamocarb (Banol) target pythium; while rhizoctonia is controlled by quintozene (Terraclor), triflumizole (Terraguard), and thiophanate methyl (Cleary's 3336, Domain). Banrot contains a combination of fungicides (etridiazole and thiophanate methyl) to provide broad-spectrum control against both pythium and rhizoctonia.

Biocontrols are becoming more widely available for controlling damping-off fungi. Some university trials have been conducted and may be helpful to growers. Growing mixes that are naturally suppressive to damping-off soil-borne fungi are also available. Test their effectiveness on just a small portion of your crop.

Prevention

Use well-drained growing media. If sand or soil is used in the mix, disinfect growing media to eliminate fungi.

Use clean equipment. While steaming is the most effective means of disinfecting, treatment with commercially available disinfectants or a 10% Clorox formulation (one part chlorine bleach to nine parts water) is also effective. Disinfectants are most effective when soil particles and plant debris have been removed from the surface to be disinfected. *Note: Disinfectants are regulated as pesticides. Contact your state regulatory agency if you have questions concerning the labeled uses.*

Circulate air to prevent wet and dry pockets.

Keep temperatures sufficiently warm so seeds germinate rapidly and seedlings are vigorous. Seedlings become more resistant to damping-off as they mature.

Manage shore flies and fungus gnats. Larvae of these insects feed on plant roots and can ingest damping-off fungi. The fungi can stay in the insect's gut from the pupal stage through the adult stage. The adult can then transmit the fungal spores aerially.

Damping-off Management

Immediately remove any diseased plants from the growing area. Don't reuse containers from diseased plants.

Identify the cause of damping-off. Don't assume a fungus is responsible for the problem—rule out environmental factors first. If a fungus is involved, identifying the specific fungus causing the disease is a must. Without an accurate diagnosis, the incorrect fungicide may be used and when damping-off isn't halted, the grower may mistakenly conclude that the fungus has become resistant to the fungicide.

Use a fungicide appropriate for the specific fungus involved. For instance, a fungicide that controls pythium won't likely control rhizoctonia. Sometimes, more than one damping-off fungus is a problem and a combination of fungicides is needed.

Apply fungicides at the rates and intervals specified on the label. You'll only get continuous effective protection when the minimum time interval listed on the fungicide label is strictly observed.

Apply drench fungicides when a crop needs watering but not when it's too dry. Wet growing media thoroughly with the full rate of fungicide.

Rotate among the available fungicides.

Pest Control, January 1997.

Controlling the Root-to-Shoot Ratio, Part I

by Dr. Roger C. Styer and Dr. Dave Koranski

Water and media quality, environment, nutrition, and moisture management all have an effect on plug growth. But some factors affect shoots more than roots and vice versa. The key to growing a quality plug is to incorporate all of these various factors into controlling the shoot:root ratio.

First, we need to define the characteristics of a quality plug:

- Proper height, with short internodes and lateral branching, if possible
- Solid green leaf color when appropriate
- Sufficient leaf expansion, with the proper number of leaves for plug size
- No buds or flowers evident for most crops
- Active, healthy root system with root hairs when visible, resulting in a readily pullable plug when moist
- No disease or insects
- Timely flowering after transplanting
- Every plug is uniform, depending on crop and plug cell size
- Tone or hardness to leaves and shoots when being shipped.

Knowing what a top-quality plug looks like is the first step in improving your own plug production. This geranium plug shows all the right attributes of a good plug.
Photo: Roger C. Styer

Evaluating Shoot and Root Growth

To control plug shoot and root growth, you must first evaluate the growth and determine if it is ahead of or behind schedule. A clear picture of the desired plug quality in each crop and in each plug size is needed, whether you're transplanting your own plugs or shipping them to other growers.

Shoot growth

Height

Most growers evaluate shoot growth by how tall the plug is above the soil line. Height can be judged by internode length, petiole length, or leaf length. For single-stem crops, such as celosia, snapdragon, and tomato, internode length is the main determinant of plug height. Tall plugs have elongated internodes, but may still have the proper number of true leaves. Crops that grow from a crown or rosette, such as pansy, cyclamen, anemone, delphinium, and ranunculus, will be judged on the petiole length, not internode length. Petiole length is controlled by the same factors as internode length. Leaf length is important for height in crops such as begonia and petunia, where there is a crown-type of growth but no long petioles. As the petunia plug gets older, however, a central stem will begin to elongate and greatly determine the height of the plug.

Leaf color

A second shoot growth criterion is the color of the leaves, which, for most crops, should be a solid green. This includes the lower leaves, where yellow color may indicate that the plugs are underfed, are stressed in some way, or have root rots such as pythium or thielaviopsis. The lowest leaves may just be shaded too much, turn yellow, and drop off (as with geranium and salvia). Dark green leaves, on the other hand, may indicate too much ammonium (NH_4) fertilizer. Pale green leaves may indicate a lack of nitrogen (N), NH_4 fertilizer, NH_4 toxicity, or a lack of magnesium (Mg).

Some crops don't produce uniformly green leaves, or at least it's not desirable that they do, being known for their colored leaves, which are generally dependent on anthocyanin pigmentation. Dark-leaf begonias should have dark, not pale-colored leaves. Coleus and hypoestes have variegated or spotted pigmentation of different colors in addition to green. Dusty miller should have silvery dust on the youngest leaves when plugs are produced in 406 or larger plug cells. Lack of colored pigmentation on these crops may indicate too much NH_4 fertilizer or a lack of stress during growing (which should come from moisture, light, or temperature).

Leaf size or expansion

Another key way to evaluate shoot growth is by leaf size or expansion. Leaves should be properly expanded for the particular crop. For many crops, the leaves should completely cover the plug tray before transplanting or shipping. Small leaf size will cause the plug tray to look sparse. Customers who receive such plug trays may think there are too few usable plugs or the plugs are too small. Plug customers will definitely complain if begonia plugs have small leaves, but will rarely complain if they have large leaves, unless using an automatic transplanter. Small leaf size may be caused by not enough NH_4 fertilizer, too much chemical growth regulator, or too intense light.

On the other hand, large leaves that are thin cause the plug tray to look overgrown and too tall. These soft leaves are particularly prone to diseases such as botrytis and leaf spots. Large, soft leaves will also make the plug more susceptible to damage during shipping and transplanting and may adversely affect growth after transplanting, especially in vegetable plugs transplanted into the field.

Number of true leaves

An indication of the plug's physiological age, the number of true leaves is a direct result of the leaf-unfolding rate. Crops grown too cool will have fewer true leaves than they should. Too many true leaves, though, may indicate the plug is too old, has been grown too warm, or has been grown with too much NH_4 fertilizer, resulting in a tall plug. The number of true leaves will also depend on the plug cell size. For example, to be considered of sufficient age and size to be shipped and transplanted, pansies may have three true leaves when grown in an 800 tray, four true leaves in a 406 tray, five true leaves in a 288 tray, and six true leaves in a 128 tray.

Buds or blooms

The appearance of buds or blooms on the shoots isn't generally desirable in a plug tray. Budding or blooming

Buds or flowers on plugs, such as these celosia, are a sign that the plugs are too old or have been stressed.
Photo: Roger C. Styer

is usually a sign that the plugs are too old or have been stressed too much. For single-stem, single-flower crops—such as petunia, celosia, French marigold, salvia, and zinnia—the appearance of buds and blooms in a plug tray will mean delayed vegetative growth after transplanting, thus resulting in delays in finished production or an undesirable finished product. The plant will put all of its energy into that main flower, and only later will it put energy into more vegetative growth (branching). For crops that are multiple-branched or multiple-plants per cell, such as alyssum, portulaca, and impatiens, plug tray budding or blooming doesn't present problems after transplanting, as the plants will continue to grow vegetatively and flower at the same time. In fact, many plug customers want their impatiens plugs to be branched and budded when they transplant, as the crop will finish faster with more blooms.

Root growth

To evaluate root growth, you must first pull up, or try to pull up, plugs to look at the roots. This is usually more difficult than looking at the tops of the plug trays to determine shoot growth. In evaluating root growth, a grower should look for plug pullability, root amount and location, root hairs and root thickness, and root rot.

Pullability

For a plug to be usable for transplanting, it needs to be pullable from the tray when moist. This means that the root system is developed enough to allow the plug media in the cell to be pulled intact and transplanted together with the roots. Otherwise, a grower will be transplanting a bare-rooted seedling. Generally, all plugs should be pullable about one week before transplanting or shipping. If not, then the root growth is behind schedule.

Root amount and location

The amount and location of the roots give you an idea of the effectiveness of your moisture management, nutrition, and environmental-control programs. Roots located mainly in the top half of the plug cell can be a result of frequent, light waterings, with the bottom half staying too dry. On the other hand, the bottom half may be staying too wet, and only the top half dries out enough to support roots. Either way, the plug will not be pullable on time.

Roots will mainly be located on the outside of the media and at the bottom of the cell, less so through the center of the cell. This is due to the fact that more air is available at the interfaces between the plastic sides of the cell and the plug media and at the bottom of the cell where the hole allows drainage and evaporation. Some crops, such as ranunculus, send down a thickened taproot, which winds around at the bottom of the cell; only later will roots come up into the rest of the plug cell.

Root hairs and root thickness

Located mainly on the outside and bottom of the plug cell, long thin roots, often called hydroponic or water roots, indicate overwatering or a plug media that has very little air porosity. Generally, these roots also lack root hairs, the fuzziness on the roots. Root hairs are necessary for expanding the root surface area, which improves uptake of the water and nutrients needed for plant growth. The plug system greatly enhances the seedling's ability to produce roots with root hairs! Bare-rooted seedlings from seedling flats do not produce any root hairs at all. Root hair development on such crops as pansy, petunia, and impatiens indicates good moisture management. There is enough water available for growth, but enough air moving into the plug media pore spaces for roots to thrive.

Once root hairs are produced, they can be lost due to high salts or overdrying, however. Root hair loss will stunt growth, delay take-off of root growth after transplanting, and may open up the root system to root rots such as pythium or thielaviopsis.

From the book Plug & Transplant Production, A Grower's Guide, *published by Ball Publishing, copyright 1997.*

July 1998.

Controlling the Shoot-to-Root Ratio, Part II

by Dr. Roger C. Styer and Dr. Dave Koranski

To obtain the proper shoot:root ratio in any plug, we first need to review and tie together the effects of water quality, media quality, environment, nutrition, and moisture management on shoots and roots. Sometimes a plug crop grows too fast, and you need to slow it down. At other times, you need to push the crop to get it ready to transplant. When shoot growth is ahead of root growth, how do you get the roots to catch up? When the plug height is small but the roots are great, how do you get the shoots to expand?

Knowing how to evaluate your crops for quality is the key to managing your shoot-to-root ratio. Here are the characteristics you should look for:

- Proper height, with short internodes and lateral branching, if possible
- Solid green leaf color when appropriate
- Sufficient leaf expansion, with the proper number of leaves for plug size
- No buds or flowers evident for most crops
- Active, healthy root system with root hairs when visible, resulting in readily pullable plug when moist
- No disease or insects
- Timely flowering after transplanting
- Uniform plugs, depending on crop and plug cell size
- Tone or hardness to leaves and shoots during shipping.

How to Adjust the Shoot:Root Ratio

Because growing a young seedling in a plug system is a dynamic process, it's difficult to keep the desired ratio between shoots and roots throughout the plug crop every time. Thus, you'll always be making adjustments for every

These pansy plugs are starting to grow too tall, as determined by petiole length.
Photo: Roger C. Styer

crop to promote the shoots or roots or to speed up or slow down plug growth, based on the production schedule.

Too much shoot growth

The main problem most growers have with plug quality is a high shoot:root ratio, or too much shoot growth. Symptoms include tall, stretched plugs; large, soft leaves; and poor root growth. Petunia and impatiens frequently exhibit this problem. Many plug growers have achieved luxuriant top growth but poor rooting, regardless of the plug tray size. Looking at the shoot growth, the plugs seem like they should be ready to transplant but they aren't pullable, due to less than desirable root growth.

Corrective measures for a high shoot:root ratio include: reducing temperature or using negative DIF; reducing moisture levels; changing fertilizer to one with more NO_3 and calcium; increasing light levels; and using chemical growth regulators. Growth of shoots and roots increases linearly with increasing temperatures until an optimal temperature is reached.

The average daily temperature (ADT) determines the leaf-unfolding rate when temperatures are maintained in the linear range for growth. To slow down shoot growth, reduce the ADT—but stay within the linear range (usually 50° to 80°F), or growth will stop completely. However, root growth will also slow down with a lower ADT. Internode length in most crops mainly depends on how temperature is delivered during a day-night cycle or on the difference between day and night temperatures. A negative DIF (day cooler than night) will keep internodes short while allowing roots to grow.

A good alternating wet-dry moisture cycle (creating moisture stress) will promote roots while keeping shoot growth from being too fast. If your plug medium isn't drying out within two days during the winter or every day during the spring, then evaluate its water-holding capacity and air porosity. It may be that the plug media contains too many fine particles with not enough air spaces for good root growth and good drainage. Media compaction through handling and stacking filled plug trays on top of each other will also squeeze air spaces out of plug media. Using watering nozzles that deliver heavy streams of moisture may compact media when plug seedlings are very young. Overwatering plugs is the most common mistake plug growers make, particularly with smaller plug sizes. Provide air movement and dehumidification cycles to promote evaporation and transpiration.

Changing to a high NO_3 and low NH_4 fertilizer with at least 6% calcium (still not enough if there is low calcium in water) will favor root growth while toning shoots. Calcium is needed for cell wall thickness and root cell division and elongation in roots. High NO_3 will keep plants from stretching, but you should still supply enough nitrogen to maintain growth. It may also be necessary to reduce the total nitrogen used to keep shoots from stretching. This NO_3 fertilizer can also be used when growing plugs at soil temperatures less than 60° to 62°F. Examples of high NO_3 and high calcium fertilizers include 15-0-15, 14-0-14, and 13-2-13-6-3. Remember that some of these fertilizers are basic and will raise media pH over time.

Increasing light levels or light intensity will increase photosynthesis, thereby providing more sugars (carbohydrates) to the roots. At low light levels (less than 1,500 foot-candles), leaves have first priority on the carbohydrates. At higher light levels (1,500 to 3,000 foot-candles), roots get their fair share of the needed sugars for growth. Light intensity greater than 3,000 foot-candles on the leaf will cause its temperature to go above the threshold (above 90°F) for photosynthesis to continue safely. The stomates in the leaves then close to protect the plant, resulting in a temporary growth regulation. When leaf temperature is back in the safe range, stomates open, and photosynthesis continues. Also, higher light levels will tend to dry out plug media faster through evaporation, and transpiration is increased, resulting in more plant uptake of calcium.

Light quality also affects plant appearance. The ratio of red to far-red light allows plants to determine whether they're being shaded by surrounding plants. Green leaves preferentially absorb red light compared to far-red light, but leaves shaded by other leaves are exposed to less red light than far-red light. A high level of far-red light stimulates stem elongation and reduces branching, enabling a plant to grow through a canopy shading it to compete effectively with adjacent plants. Plant crowding is common in a plug tray. Overlapping leaves filter out red light, resulting in greater stem elongation and less branching in lower internodes. Hanging baskets over a plug crop will promote the same effect.

Chemical growth regulators are traditionally used when shoot growth is ahead of root growth. These chemicals act to block the synthesis or activation of giberellic acid (GA), which would promote internode elongation. Some chemicals, such as Bonzi and Sumagic, also stimulate root growth while keeping shoot growth from expanding, particularly in impatiens.

Slowing down shoot growth while simultaneously speeding up root growth isn't an overnight process. It may take a week or longer to get enough roots, depending on the conditions available for manipulation and the crop being grown. This process is different from just applying a chemical growth regulator and stopping the shoot growth within a day. Controlling growth through moisture and feed consists of decisions that should be made wisely each day for each crop.

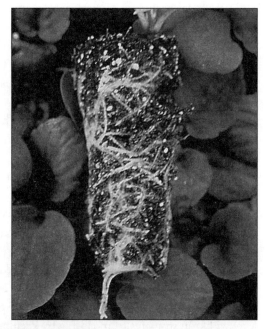

Too much root growth

In some cases, plugs may have very good root systems, but the shoot growth is too small for the age of the plug. This results in a low shoot:root ratio. Symptoms of root growth being ahead of shoot growth include small and light colored leaves, small tops with very

Even distribution, plenty of root hairs, and a firm root ball are all signs of a good root system.
Photo: Roger C. Styer

short internodes, and numerous roots, making the plug very pullable. Commonly, growers in warm areas, such as California and Israel, with low humidity and high light levels may experience this problem.

Corrective measures for promoting more shoot growth include increasing temperature or using positive DIF; increasing moisture levels; increasing use of fertilizers containing more NH_4 and phosphorus and less NO_3 and calcium; reducing light levels; increasing humidity levels around plugs; and minimizing use of chemical growth regulators. Shoot and root growth increases linearly with increasing temperatures until an optimal temperature is reached. Plant growth decreases above that temperature. ADT determines the leaf-unfolding rate when temperatures are maintained in the linear range for growth (usually 50° to 80°F). To speed up shoot growth, a grower can increase the ADT within the linear range until an optimal temperature is reached. A positive DIF (day warmer than night) will increase internode length, or stretch, and plant dry weight.

To keep plug cells from drying out too much, water more frequently and more thoroughly. However, you don't want to lose the good roots you've gained by keeping the media moist all of the time. Too much moisture for too long will promote foliar diseases (botrytis, leaf spots) and root rots (pythium, rhizoctonia). In climates that tend to be very dry with high light levels, plug growers should evaluate plug media with a higher water-holding capacity and less air porosity.

Fertilizers higher in NH_4/urea and phosphorus and lower in NO_3 and calcium will promote rapid shoot growth. Plants growing in conditions with warm temperatures, high light levels, and dry air are working very hard, with maximum photosynthesis and respiration, and need more energy. More NH_4/urea and phosphorus are needed to expand the leaves and internodes and keep the green color. Less calcium is needed, with the plant actively taking up available calcium through rapid water uptake. Higher rates of total nitrogen (ppm) may also be required.

To expand shoot growth, use shade to reduce light levels to less than 2,500 foot-candles. A high light level can act as a growth retardant by heating up leaf temperatures and shutting down photosynthesis. A lower light level will keep stomates from closing by keeping leaf temperatures below the danger level (less than 90°F). Evaporation and transpiration will also be slowed down under lower light levels. Providing more far-red than red light will also promote rapid shoot growth.

Increasing humidity levels around plugs can be accomplished by maintaining higher moisture levels or by introducing fog intermittently through the greenhouse. Fog systems can also cool the greenhouse or aid in the germination process on the bench. Pad and fan cooling also add humidity to dry greenhouses. High humidity levels reduce transpiration or water loss from the plants and reduce evaporation of media water.

If you're growing in warm, dry areas with high light levels, reduce chemical growth regulator use to only what's really necessary. For many crops, such environmental conditions will be enough to keep shoot growth from rapidly expanding and keep internodes short. All chemical growth retardants should be closely trialed so as not to overdose a whole crop of seedlings. However, when the weather turns cloudy or rainy for several days, you may need to use such chemicals to keep plugs from rapidly stretching.

Generally, if root growth is sufficient but shoot growth lags behind, it's easier to speed up shoot growth by using positive DIF, feeding with more NH_4, cutting down light levels, and keeping more moisture in and around plugs. Within three days, shoot growth increases should be evident. Many plug growers hold back on feed levels, particularly with vegetables, until about three days before shipping and planting. Then, they pump up seedlings with 300 to 400 ppm nitrogen from fertilizers such as 20-10-20 to green them up and get them growing again. This technique works well, when environmental conditions cooperate with the desired shoot growth. However, under conditions of warm temperature, high light, and dry air, it may take longer than three days for shoot growth to expand sufficiently from fertilizer alone.

Behind schedule

When a plug crop is behind schedule and overall seedling growth (roots and shoots) needs to be speeded up, you can:
- Increase the ADT by 5°F.
- Use positive DIF.
- Practice a good alternating wet-dry watering cycle.
- Increase feed levels to 150 to 250 ppm nitrogen, using more NH_4/urea fertilizer (such as 20-10-20) and feeding more frequently.
- Provide light levels of 1,500 to 2,500 foot-candles. Keep a close eye on media EC and pH levels so as not to run into any major problems with root growth or nutrient uptake.

Ahead of schedule

When a plug crop is ahead of schedule and overall growth (roots and shoots) needs to be slowed down, you can:
- Decrease ADT by 5°F or more.
- Use negative DIF of 5° to 10°F.
- Run plugs drier before watering.
- Decrease feed levels to under 100 ppm nitrogen using more NO_3 and calcium fertilizers (such as 14-0-14 or 13-2-13-6-3).
- Increase light levels to greater than 2,500 foot-candles but less than 4,000 foot-candles.
- Use chemical growth regulators judiciously. Late applications of Bonzi or Sumagic may cause delays in plugs growing out after transplanting.

Again, monitor media pH and EC levels, particularly when using the basic fertilizers just mentioned and drying plugs out more thoroughly. Soluble salt levels around roots will increase three to four times under dry conditions.

Key Points to Remember
- Plug quality is determined by the shoot:root ratio, along with leaf color and other attributes.
- Learn to recognize what makes a good shoot or good root system for each type of plug crop you grow.
- Factor levels such as light, moisture, and fertilizer that promote shoot growth may be the opposite of those that promote root growth.
- To speed up shoot growth, increase temperatures up to the optimal level or use positive DIF, increase moisture levels, increase feeding with NH_4 and phosphorus, and decrease light levels.
- To speed up root growth, use negative DIF, increase light levels, increase feeding with NO_3 and calcium, and decrease moisture levels.

From the book Plug & Transplant Production, A Grower's Guide, *published by Ball Publishing, copyright 1997.*

August 1998.

Key Tips for Bench-Top Germination
by Dr. Roger Styer and Dr. Dave Koranski

A portion of a greenhouse can be devoted to plug germination if the proper conditions are maintained. Greenhouse germination allows easy access to, and close inspection of, the plug trays. Root-zone heating, with assistance from a perimeter or overhead heating system, should be set up to provide optimum soil temperatures. In warm weather, a cooling system will reduce soil temperature to the proper range for germination. An overhead fog or fine mist system can supply the necessary moisture. Irrigation booms can be used if very fine nozzles (50 to 80 microns) are installed. Tempered water can also maintain soil temperatures, especially if cold well water is used during the winter season; heat the water to 70°F. Moisture can be controlled with timers (fog or mist systems) or according to visual inspection (boom irrigators or hand-watering).

Some growers use capillary mats when germinating plugs on benches, especially if using root-zone heating without fog. The mats supply moisture and humidity to the plug trays and reduce the drying effects of the root-zone heating. The mats also help distribute heat evenly across the bench and help pull excess moisture through the plugs.

The disadvantages of capillary mats are: (1) roots may grow into them; (2) buildup of algae, fungus gnats, and shoreflies; and (3) plug trays may actually stay too wet. Capillary mats are not as widely used in the United States as they previously were.

In northern areas during winter, HID lights are recommended to provide supplemental light energy (about 400 foot-candles at plant level) for seedling

The more control you have over temperature and moisture, the better you bench-top germination results will be. C. Raker and Sons, Litchfield, Michigan (pictured), germinates nearly all of their 100 million or so plugs on benches.

development after germination. In the greenhouse, HID lights can provide both light and heat, raising the soil temperature 2° to 5°F. In sunny weather the greenhouse germination area should be shaded to provide less than 1,500 foot-candles.

A microenvironment can be created for the seeds during Stages 1 and 2 by covering the plug trays. Coarse vermiculite, plug media, sand, and other materials have been used successfully by many growers. Other types of coverings can trap heat and humidity around the seeds, but they are then removed once germination is visible. These coverings include polyethylene film (clear and opaque), Vispore (plastic punched with holes), and frost covers (spun porous plastic, sometimes called Agricloth). Water can be applied to the plug trays through Vispore and frost covers, as long as someone watches carefully for water accumulation or dry spots. All of these plastic products can be cleaned off and reused.

Using the greenhouse bench for germination requires careful grouping of crops by temperature and moisture requirements. Dividing this greenhouse area into temperature zones gives you more flexibility for placing crops in the best germination climates. How long a crop remains there depends on the temperature and moisture conditions, type of crop, variety, and seed quality. Once germination is visible, moisture and temperature may need to be lowered slightly to keep seedlings from stretching. If this is not possible in the germination area, then trays will need to be moved to another part of the greenhouse for the rest of Stage 2 through Stage 4.

Key Points to Remember

1. Germination depends heavily on your ability to optimize and control soil temperature and moisture at the seed level for each specific crop. Germination chambers give you better control of temperature, moisture, and light, but they are more costly to build. Greenhouse benches make it easy to see germination, but a grower needs to pay more attention to controlling temperature and moisture.

2. Space requirements can play a role in deciding what type of germination facility to use, especially during peak times of the year. Germination chambers will free up bench space for up to seven days, depending on the crop.

3. Chamber-building costs should be determined through careful planning. Many growers build their own and improve them along the way.

4. The type of germination facility you use will largely be determined by what types of crops you grow and at what time of year. Growers in the southern United States cannot germinate pansies very well on the greenhouse bench in July and August, when temperatures are above 90°F. Vinca will have difficulties germinating in the northern United States during January and February, when greenhouse temperatures are below 70°F.

5. Finally, attention to detail is essential for high germination results, regardless of the type of germination facility. Be vigilant in removing trays from the germination chamber before stretching occurs. On the greenhouse bench, more attention must be paid to moisture levels, which may fluctuate rapidly, as well as soil temperature.

From the book Plug & Transplant Production, A Grower's Guide, *published by Ball Publishing, copyright 1997.*

November 1998.

Plug Deficiencies

by Paul Nelson and Jin-Sheng Huang

At North Carolina State University, we've researched deficiency symptoms of nitrogen, phosphorus, and potassium for five plug crops. The crops included begonia Prelude Mix, impatiens Hybrid White, marigold Discovery Orange, pansy Regal Mix, and petunia Primetime White. Plants were grown in 288-cell plug trays in a medium consisting of three parts sphagnum peat moss to one part perlite. We created the various deficiencies by omitting the nutrient we were studying from the preplant and postplant fertilization program.

Nitrogen

Nitrogen (N) deficiency symptoms generally follow a sequence, including uniform chlorosis (yellowing) of the entire lower leaf, followed by necrosis (dying) of the lower leaf, and finally abscission (dropping) of the lower leaf if the plant species has the facility for dropping leaves. Only impatiens have this capacity. However, this syndrome describes nitrogen deficiency in impatiens and petunia.

Begonia differ, however. The lowest leaf turns uniformly chlorotic, but leaves farther up the plant develop a red margin as each turns uniformly chlorotic with advancing nitrogen deficiency. Also, necrosis that follows chlorosis begins and remains for a long time at the terminal end of leaves.

Marigold also differs. As leaves turn uniformly chlorotic, the tips of these leaves develop a purplish red pigmentation. An additional symptom in marigold is seen in the upward positioning of leaves into a funnel shape.

Normal petunia plants are shown on the left and nitrogen deficient plants are on the right. The deficient plants are small, and the lowest leaf is turning uniformly chlorotic.
Photo: Paul Nelson and Jin-Sheng Huang

Pansy follows the traditional symptoms of nitrogen deficiency seen in petunia and impatiens, with one exception. Sometimes leaves develop a purple pigmentation as they turn uniformly chlorotic. This tends to mask the chlorotic symptom.

Phosphorus

Phosphorus (P) deficiency symptoms follow the traditional syndrome, including stunting of the plant, leaves turning a deeper green than normal, then uniform chlorosis of the lowest leaf, followed by necrosis of the oldest leaf. These symptoms progress up the plant with time.

Stunting from N and from P deficiencies is about equal in magnitude. This syndrome completely describes P deficiency in petunia. In the other crops most of this syndrome appears along with other symptoms. The lower leaves of impatiens abscise before necrosis occurs.

Marigold differs the most from the traditional symptoms. Chlorosis begins at the tips of older leaves, followed by purple pigmentation, and then

necrosis at the tip. Sometimes all three symptoms are seen in the lower leaf before the green pigmentation disappears at the base of these leaves.

Begonia symptoms are somewhat similar to marigold. Chlorosis begins at the tip of older leaves, followed by necrosis at the tip. These symptoms progress toward the leaf base with some leaves displaying a green base, chlorotic midsection, and necrotic terminal area.

Phosphorus deficiency in marigold is shown here. Notice the necrosis evident on the tip of older leaves.
Photo: Paul Nelson and Jin-Sheng Huang

The lower leaves of pansy often progress from uniform chlorosis to a uniform white color. Sometimes a purple cast develops in the lower leaves.

Potassium

Potassium (K) deficiency symptoms, as in the cases of N and P deficiencies, begin on the lower leaves. Impatiens and marigold follow the more typical symptoms. In these plants the deficiency begins with a very short-term chlorosis of the terminal edge of the lower leaf, followed quickly by necrosis of this tissue. The symptoms spread toward the base of the leaf and then upward to the next leaf. Unlike the other four crops, the leaves of marigold curl down, giving the plant an umbrella shape. At the time that chlorosis forms in impatiens plants, white spots may also form along or near the terminal leaf margin.

K deficiency starts in pansy plants as uniform chlorosis of lower leaves that gives way to necrosis, and at the same time white patches form on the midplant leaves, with these progressing to necrosis.

In addition to necrosis along the leaf margin, potassium deficiency on impatiens may originate as whitish tan spots on older leaves. Although not always, these spots generally originate near the terminal leaf margin.
Photo: Paul Nelson and Jin-Sheng Huang

The deficiency in petunia starts as small, whitish tan spots along or near the terminal margin of the oldest leaf. As these spots turn necrotic, chlorosis, followed by necrosis, develops between the spots.

Begonia symptoms are the most distinctive. A chain of whitish tan spots forms along the terminal margin of the oldest leaf. A brown ring soon forms around the margin of each spot. The spots grow larger and turn into a necrotic band. The band progresses toward the base of the leaf. Before half of the lower leaf becomes necrotic the same symptoms begin on the second leaf from the plant base. By the time the lower leaf is completely necrotic, the terminal ends of the lower three or four leaves have these same symptoms.

September 1998.

Bugs on Plugs: Fungus Gnats and Shore Flies

by Dr. Roger C. Styer and Dr. Dave Koranski

Most plug seedlings are produced in a relatively short period of time. They're smaller and don't have flowers, compared to post-transplanted crops, thus making plugs not as likely to have many problems with insects. However, with their intensive, high-density culture and the many crops that may be grown at the same time, certain insects can become major problems even during the short plug cycle. With longer term plug crops, such as lisianthus, gerbera, cyclamen, and begonia, diseases and insects have a better chance to adversely affect quality.

The main problem insects in plug production are fungus gnats and shore flies. Large numbers of adults create a nuisance and leave fly specks on leaves. More seriously, adults and larvae have been shown to transmit some fungal root rots, such as pythium. Fungus gnat larvae also feed on roots, thereby exposing them to root rot.

Even though they may be small and quick to grow, plugs can still attract pests, especially fungus gnats and shore flies.

Telling them apart is the first job in controlling them. Fungus gnat adults have long antennae and long, thin bodies and are lazy fliers, like mosquitoes. Shore fly adults have short antennae and short, stout bodies, and are active fliers, like flies. Yellow sticky cards placed close to the plug trays will allow you to see which pest you have. The fungus gnat larva has a black head, whereas the shore fly larva doesn't.

The life cycles of these two insects are similar. Eggs are laid in the soil and hatch into larvae. The larvae, found within the top inch of soil, feed on algae and fungal materials. Fungus gnat larvae feed on roots, if necessary. The larvae will pupate and can persist through drought and temperature extremes in the soil. Adults emerge when environmental conditions are suitable. The life cycle is temperature-dependent and ranges from two to four weeks. Development time decreases as the temperature increases (for example, twenty days at 75°F; fifty days at 55°F).

Both shore flies and fungus gnats need high moisture levels to develop, so conditions that favor algae and fungal growth enhance these insect problems. To control these insects, you can:

- Reduce excess water use and runoff.
- Prevent water from collecting under benches or on floors.
- Control algal development with less moisture, reduced fertilizers, and quaternary ammonium salts (Physan 20, Greenshield), or Agribrom.
- Keep floors and benches clear of plant debris and soil.
- Keep the greenhouse weed-free.

The areas under benches can be kept weed-free and inhospitable to fungus gnats and shore flies by using hydrated lime, copper sulfate, or Agribrom. Hydrated lime is mixed as a slurry at one and a half pounds per gallon of water and sprayed under benches. Copper sulfate is used at one pound per gallon of water and should be effective for three months. Both of these materials are highly phytotoxic and should never contact the crop. Avoid overuse of these chemicals, as problems can develop with potential runoff.

Most pesticide applications are directed at the larvae. Make applications as drenches or coarse sprays to the growing media surface, getting the chemical at least into the top half inch of soil. Materials reported to provide good control against larvae include Gnatrol (*Bacillus thuringiensis,* for fungus gnats only); microencapsulated diazinon (Knox-Out) or dursban (DuraGuard); oxamyl (Vydate); kinoprene (Enstar II); neem (Azatin); cyramozine (Citation); microencapsulated pyrethrin (X-clude); and parasitic nematodes (X-Gnat, for fungus gnats only). Follow label directions for repeating treatments. With insect growth regulators—such as kinoprene, neem, and cyramozine—or with parasitic nematodes, don't expect immediate kill, but repeat the applications weekly. For best control, use one chemical for two or three weeks and then rotate to another.

When populations of adults get too high, aerosols may be needed. Materials reported to provide good control of adults include cyfluthrin (Decathlon), pyrethrin (X-clude), and resmithrin (PT 1200). Repeat applications for adults weekly until the population is reduced, and be sure to combine with applications for larvae. Adult control is best obtained with a fog or aerosol. Adults will move from one part of the greenhouse to another if a spray is used.

From the book Plug & Transplant Production, A Grower's Guide, *published by Ball Publishing, copyright 1997.*

Pest Control, September 1999.

Understanding Seed Physiology
from a Grower's Viewpoint

by Dr. Daniel Cantliffe

With the advent of precision seeding and mechanical harvesting, growers and seedsmen are more demanding of good seed germination and seedling growth. However, poor-quality seed will eventually lead to sales losses by the transplant producer. Seed storage conditions can adversely affect seed longevity, germination, and seedling vigor. Storage at high temperature and high relative humidity can rapidly reduce germination and seedling vigor. All of these factors must be closely monitored and maintained to produce the highest quality plants possible, leading to the greatest economic return.

Germination, in the strict sense, is the resumption of active growth of the embryo, usually after a state of rest. This results in the breaking of the seed coat and the emergence of a young plant. A seed physiologist may measure radicle germination after measuring biochemical changes in the seed or at radicle emergence. A seed technologist measures germination only after a normal plant is observed, i.e. root and shoot that are normal in appearance. On the other hand, a farmer only counts seedlings that have emerged from the soil as germinated. This may or may not be the point where a grower counts a seed as germinated. Generally, only useable seedlings are factored-in and counted as germinated seedlings.

The Phases of Seed Germination

Germination of crop seeds follows a specific sequence of events. These events include imbibition of water, enzyme activation, initiation of embryo growth, rupture of the seed coat, and emergence of the seedling. Imbibition, or the movement of water into the seed, is termed the passive phase. This first occurs as a physical movement through natural openings in the seed coat—water moves throughout all of the seed tissues. The rate and total volume of water movement is dependent on the seed coat, seed composition, and, to a lesser extent, temperature. Seeds such as soybean, which contain protein as the major storage component, will reach a larger final volume than seeds that contain a large amount of starch, such as corn. After an initial rush of water movement into the seed there is a lag phase, where respiration starts and the imbibition rate slows down. After this phase, water movement begins again due to growth, usually of the radicle through the metabolic processes going on in the seed. This is more related to an active phase of water uptake because it represents active seedling growth.

Water causes the cells in the seed to become turgid, the entire seed enlarges in volume, and the seed coat becomes more permeable to gases such as oxygen and carbon dioxide. As the seed swells, the seed coat may rupture, facilitating water and gas movement. In general, dry seed moisture content is from 5 to 8% on a fresh weight basis. Imbibitional seed moisture content will rise rapidly to over 60 to 80%. The embryonic axis will have to attain a moisture content in excess of 90% for radicle development, whereas, other portions of the seed may still be below 50% moisture after twelve hours of imbibition. This is particularly true for starchy seeds.

These are the three stages of water uptake. Phase I may be the most rapid, usually lasting from one to eight hours. Lettuce seeds will generally complete Phase I imbibition in one to two hours. So long as the seed coat does not interfere with water uptake, imbibition in Phase I is similar in both dead and living seeds as well as dormant and non-dormant seeds. This is why it is termed passive.

Phase II, or the lag phase, can last from several hours to several days, and longer if the seeds are dormant. Phase II generally concludes when the radicle protrudes through the seed coat. It is during Phase II that the major metabolic events take place which lead to the completion of germination. Dormant seeds remain in Phase II until dormancy is released. Phase II metabolic events include membrane reorganization, enzyme activation, protein synthesis, storage material breakdown, RNA synthesis, and sugar metabolism for energy derivation. Many times dormant seeds will have elevated levels of respiratory activity during Phase II, as well as certain types of synthetic processes taking place. However, depending on the type of dormancy, these seeds don't usually begin cell division. Most transplant crops don't go into a dormancy phase, although some flower crops may (i.e. geranium, which possesses a hard seed coat that can restrict water uptake).

The final phase, Phase III, is also a period of rapid water uptake. This is generally related to cell division and cell expansion, radicle protrusion, and eventually hypocotyl elongation and protrusion from the seed coat. This marks the end of germination and the beginning of seedling growth.

Factors Affecting Germination

The length of time for germination and its various phases is dependent on many factors. In the transplant house, it is dependent upon moisture availability, media composition, aeration, temperature, and light, when needed. Under conditions of high or low temperature, the processes of Phase II are much delayed. Seeds that have been planted in moist media should be maintained close to their optimum germination temperature. This temperature will provide the most rapid and uniform emergence of seedlings.

Transplant plug growers should also understand that various seeds respond differently to the above-mentioned environmental conditions, and that the quality and vigor of the seed itself will predispose the seedling to its optimal growth rate. Such things as the size of the seed, the composition of the seed, the size of the embryo, and seed coat permeability (in regards to water and gas exchange) all influence the rate of imbibition in Phase I and the length of time that it remains in Phase II. Generally, Phase I is more affected by water availability and inherent seed characteristics than other outside environmental conditions.

Moisture

For plug producers, it is generally easier to maintain adequate moisture levels in the transplant tray than it is to control temperature. Wet seeds immediately following planting to offset the problems of maintaining seeds in trays or in field conditions wherein soil moisture conditions are variable, and thus, initiating germination in the seed population at different times. If seeds are not wetted immediately after planting, certain seeds in a population may be predisposed to moisture levels that will initiate the early stages of germination. Thus, when the entire population is then wetted, it can lead to variable stand establishment.

Some flower seeds need wet conditions (i.e. continuous field capacity). These seeds include impatiens, begonia, pansy, ranunculus, and cyclamen. Most vegetable seeds germinate best at field capacity. Some seeds germinate under dry conditions. This means high humidity without excessive moisture. These seeds include seedless watermelon, aster, zinnia, verbena, and delphinium. Excessive moisture can lead to anaerobic conditions, especially when media and soil types are quite dense. Stacking flats in the germination room further exacerbates this condition. Conditions of too much moisture can reduce germination and make the seeds more sensitive to high temperatures. Thus, moisture, temperature, and gas exchange are intimately interrelated with one another.

Temperature

Temperature can be difficult to control, especially under summer conditions. In the greenhouse, placing the transplant trays in a controlled-temperature room may best control germination. The period of time in the germination room should last no longer than to the initiation of the radicle protrusion.

A limiting step in germination may occur any time temperatures fall below or above the optimum for germination of any particular seed. Temperature, especially high temperature, can lead to reduction or inhibition of germination in many seeds. This is especially true of several vegetable and ornamental flower seeds when the temperature rises above 86°F for extended periods of time. In many of these seeds, this inhibition of germination only lasts as long as the seeds are exposed to the excessively high temperature. Thus, as temperature is reduced towards the germination optimum, the seeds will usually germinate.

Many times, reduced night temperatures are not long enough at an adequate temperature to allow germination to occur. Unfortunately, because of several factors including seed quality (vigor), moisture availability, and variability in temperature within a tray, the plug producer may once again find that seed lots may vary greatly in their emergence capacity. In some seeds such as lettuce, geranium, and impatiens, imposing a high temperature for periods in excess of approximately seventy-two hours may impose a secondary dormancy called thermodormancy. Geranium and impatiens are inhibited by temperatures in excess of 77°F, and many, if not most, lettuce cultivars are inhibited by temperatures of 86°F or more.

Aeration

Adequate aeration is another factor that must be considered to ensure optimum and uniform germination. Transplant plug producers who use germination rooms often don't consider this factor when trying to establish uniform emergence. In these cases, the trays are filled with media and moistened with water then stacked one atop the other to the point that aeration can be limited, especially in the trays anywhere below the top of the stack. Sometimes these stacks will be on a palette from floor to ceiling in excess of ten feet high. Many plug producers have developed tray styles and stacking procedures, which allow aeration at the seed level in each of the trays.

It is generally necessary to cover many seed species during germination with media that will maintain high humidity and moisture, but allow the maximum aeration at the seed level. Such materials as coarse vermiculite, sand, perlite, Styrofoam beads, or calcine clay can be used to cover the seed. Smaller seeds shouldn't be planted and covered as deep as large seeds.

Light

Many crop species don't require light for germination. However, for some crop species, for example tobacco, light is a requirement. Light can actually inhibit radicle extension and growth in some crops, thus inducing non-uniform seedling growth. This is evident in such species as vinca, cyclamen, phlox, and lettuce. In other species, germination may be improved through the addition of light. These crops include impatiens, petunia, and celery.

Seed Testing and Vigor

A seed-testing laboratory can help determine the ability of a seed to grow by reporting percentage of germination. Most testing labs will test seed for purity, moisture content, and percent germination. Unfortunately, seed labs cannot give an accurate idea of how seeds will perform under varying field conditions. They only report how well the seeds germinate under ideal conditions. Testing labs often use paper towel tests, which provide the seed with continued optimum moisture conditions. The seeds are then placed in a controlled-temperature germination chamber that meets that particular species' optimum conditions. Thus, the results of the germination test may be grossly misleading to a transplant producer when crops are grown under environmental extremes.

Seed-testing labs generally do not provide tests for seed vigor. Vigor can be defined as the ability of a seed to germinate rapidly and produce a normal seedling under a wide range of conditions. Seed vigor is something that cannot be measured until the seed germinates. Even then, vigor measurements are hard to correlate to final yield. There is no universal vigor test for all seeds. A vigor test can only measure one phase of early seedling growth. Plug or field producers of transplants should test and record germination across all conditions an entire seed lot will experience in order to determine seed uniformity and total emergence under those varying conditions. This, in most instances, gives a good indication of the potential vigor of a particular seed lot.

There are several tests that are used by seed companies to measure vigor in the seed lot. Several examples of tests for seed vigor include: looking at either cool and/or warm germination stress tests; uniformity and rate of radicle protrusion; measuring radicle growth over a certain period of time; conductivity of seed leachate; accelerated aging; various seedling growth tests, such as root length and seedling height; and, more recently, a technique developed by the Ball Seed Company utilizing image analysis of cotyledon expansion.

Vigor tests are often used by a seed company to determine which seed lots are strongest and, in some instances, predict how long a seed lot will store. Seed companies can also use vigor tests to determine potential markets in which the lot should be sold. Companies using such tests will generally direct their best seeds to markets that require and will pay for high-quality seeds. The grower who purchases seed from these companies will generally find that total stand counts and emergence uniformity are improved. Seed companies will often charge more for higher quality lots. If poor seed must be removed from a lot, this will increase the price of the remaining seed.

A transplant plug grower essentially is selling rental space in a transplant house. The use of the highest quality seed (highest vigor) will help ensure rapid, uniform, and optimum stands of the crop being grown. This translates into greater profits for the producer. High seed vigor can improve the rate and uniformity of germination and the rate of early seedling growth, especially as it translates to plant growth under less-than-optimum conditions.

Frequently, the seed viability (ability to germinate) and seed vigor are directly related. Both viability and vigor decline with time. Generally, vigor begins to decline before the producer observes a decline in viability. This means that a seed lot that germinates uniformly at 90% may be adversely affected as environmental conditions become more stressful. If a grower uses this seed lot in later plantings, vigor may be decreased. In this case, the seed lot might germinate well (90%) at conditions close to optimum, but as stressful environmental conditions ensue, the seed lot may fail to germinate or become non-uniform in its germination pattern.

Seed Storage

The process of aging occurs naturally during storage but can be artificially accelerated by high temperature and high relative humidity during storage. Optimum seed moisture contents during storage for many crop species is in a range on 5 to 8%. If moisture content drops below the 5% level, storage life and especially seed vigor may be decreased. This is generally related to a disruption of membrane organization and is irreversible. When seed moisture contents go above 12%, various fungi and insects can grow and reproduce in and on the seed. At this moisture level, aging is accelerated because the processes of Phase II imbibition begin but cell division and elongation can't occur. Thus, seed storage conditions are of prime importance to maintain seed viability and seed vigor, processes that can then relate to seed longevity. Good storage conditions for most seeds are between 41° and 50°F at approximately 40 to 50% relative humidity. If relative humidity can be controlled to 30%, seed longevity can be further improved.

Seed companies will store and ship many types of seeds in hermetically sealed containers. This can be in foil packets, cans, or plastic containers. Such sealed containers provide an excellent barrier to moisture movement in and out of seeds. Unfortunately, these storage containers do not maintain a barrier against temperature. Plug producers are cautioned in all stages of seed movement—from the seed companies to the plug tray—to try to maintain optimum temperature conditions. Once these containers are opened and the moisture barriers are removed, the plug producer will then have to maintain adequate humidity levels in the storage area in order to maintain seed longevity over an extended period of time.

Many large plug producers have temperature- and humidity-controlled seed rooms with alarm systems that indicate if either the temperature or humidity deviates from the desired range. For smaller plug producers, the use of a frost-free refrigerator will substitute for a high-tech seed room. When in the refrigerator, seeds should be placed inside a resealable container. When seeds are needed for use, the container should be brought out and kept closed until the container temperature has equilibrated to room temperature. This prevents moisture condensation inside the container and on the seeds. Once the seeds are used, any remaining seeds should once again be tightly sealed within the container and placed back into the refrigerator. Seeds should be stored in waterproof and vapor-proof containers, such as Tupperware containers, Mason jars, or five-gallon or smaller plastic cans with sealable lids. A layer of silica gel can be placed at the bottom of the container, and seeds should always be equilibrated to room temperature before storage containers are opened. If the seed is to be used over an eight-hour or longer period, seeds should be retained in their containers, and only the amounts needed during the time of use should be removed.

If the plug producer suspects that seeds are gaining moisture through the process of moving in and out of the refrigerator, then the moisture content of the seeds should once again be brought down to the 5 to 8% level. This can be achieved by placing the open container in a room at low relative humidity or by using a thin layer of silica gel as a desiccant in the bottom of the storage container. Seed storage containers should be equilibrated in an air-conditioned room or an area with lower relative humidity.

Some plug producers set up a seed room as part of a controlled-climate room or walk-in cooler. Home-use dehumidifiers are not suitable for controlling moisture content in these rooms, because they will freeze up at temperatures below 63°F and add heat into the room. Germination should be routinely checked for seeds that are stored for periods of six months or more. The plug operator should determine whether or not a seed lot should be retained if germination and/or the ability to germinate under stressful conditions become a problem.

Seed Pathogens, Pelleting, and Enhancements

Plug producers should be continually aware of the problems related to seed-borne pathogens in reducing stand and in potentially spreading the pathogens throughout the transplant production area. For this reason, using seeds that have been treated with various fungicides labeled for that particular seed species is recommended. Further, the use of film coating will ensure safe conditions for the transplant operator by reducing chemical dust in the atmosphere.

Where high-volume precision placement of the seed is required, pelleted seed is recommended. The coatings used in pelleting are usually some type of clay mixed with a binder and/or water. The plug producer should only purchase pelletized seed that has been coated recently. Seed storage conditions for pelletized seeds are the same as for those regular seeds. In some instances, and for some species, pelletized seeds may not last as long in storage as conventional seeds.

Certain seed enhancement treatments can improve the rate and uniformity of germination, especially under less-than-ideal conditions. Many seed firms are selling primed seed. Seed priming refers to hydrating a seed under controlled conditions, permitting the initial germination process to begin, while preventing the radicle from emerging through the seed coat. During priming, the moisture content may increase to 40 or 50%. Cell division and/or cell elongation doesn't generally take place during the priming process. Most importantly, after priming, the moisture in the seeds is reduced to the initial content of between 5 and 8%. Primed seeds can then be packaged as normal nonprimed seed and be shipped and planted using conventional seeding equipment.

Unfortunately, seed storage time is often reduced in primed seeds. Also, seed priming increases the rate of germination and generally the uniformity of germination under a wider range of conditions, but doesn't increase total germination under ideal conditions. In other words, seed priming can't make dead seeds come alive. The variables, which are controlled in priming, include the amount and the rate of water uptake, temperature, and duration of the process. Seeds, which require light to germinate, should be given light during the priming process. Oxygen should be made available to the seeds at all times, because the seed is a living entity and requires oxygen for respiration.

Understanding seed germination is extremely important to the plug producer. Optimizing seed germination is the most essential step in helping ensure economic returns in the plug operation. Germination dictates final stands that the plug operator will achieve, but also the uniformity of emergence will ensure a high-quality plug crop. The use of high-quality seeds and proper, careful maintenance of the seeds' environment will help optimize these processes to produce a high-quality plug crop.

From a presentation at GrowerTalks' 1998 International Plug Conference, October 8-10, 1998, in Kissimmee, Florida.

Chapter 4

Taking Production into the 21st Century: Equipment & Automation

---

Setting up Your "Dream" Seeding Room

by Dr. Roger C. Styer

Now that you've made it through another busy spring season, it's time to gather your thoughts and turn them toward making improvements for next year. In reviewing your plug production, were you able to get all of your plug trays sown accurately, efficiently, and on time? Did you need too many people filling plug trays, applying labels, sowing seeds, covering, and watering them in? Did your equipment have trouble handling large or odd-shaped seeds? If you could redesign your seeding room, how would you do it? The answers to these and other questions can be solved by having the proper equipment and by setting it up so it works as a system.

What Are Growers Currently Using?

For small and medium-sized plug growers, seeding equipment fulfills their basic needs. Most growers have a flat filler capable of filling plug trays and finished cell packs. They dibble the trays by hand before seeding or with a simple roller dibble. They have one seeder that must meet all of their sowing needs. Some growers have a topcoater, but many manually cover trays with media, coarse vermiculite, or plastic. They probably use a watering tunnel, which doubles for watering in bedding flats. Filling, seeding, covering, and watering are done separately, usually by the same person or by part-time help.

Traditionally, these growers have used template seeders or manifold vacuum-tip seeders. One person is responsible for all seeding. Odd-shaped seeds are sown with the same seeder, although rather slowly and with difficulty.

Today, these growers are starting to buy needle seeders, which can sow large or odd-shaped seeds better and faster. As they sow more plug trays, they replace their template seeders with faster seeders, such as manifold vacuum-tip seeders. If they have more money to spend and their seeding needs increase, they move up to high-speed drum seeders.

For large plug growers, seeding equipment must be accurate, fast, flexible, and integrated (able to work together as a dedicated system). These growers already have the essential seeding equipment, but are looking to integrate their seeding lines to reduce labor and speed up the seeding process. Their flat fillers are set up to only handle plug flats. Tray labeling is still done mostly by hand. More attention has been paid to dibbling the trays before seeding, topcoating automatically with media or coarse vermiculite, and using a dedicated watering tunnel to water the trays before putting them into a germination chamber or on the bench.

Generally, these large plug growers have more than one seeder to handle their high plug volume and likely have more than one type of seeder to sow different tray sizes or odd-shaped seeds. Due to their seeding volumes, these growers tend to have drum seeders (sometimes called cylinder seeders). With speeds up to 1,200 trays per hour, it's easier to get through heavy sowing weeks with these seeders. Some drum seeders work using water pressure inside the drum; nearly all now use air pressure to hold and distribute the seed.

Thanks to improvements in technology, modern seeders are quieter than previous models, take up less room, are easier to use, require less seed, and cost less to purchase and operate.

Manufacturers say drum seeder sales are strong, but beginning to taper off. However, needle seeder sales are increasing dramatically. Many large plug growers are using them to sow odd-shaped seed, such as marigold, zinnia, and dahlia. Needle seeders are also approaching the speed of drum seeders.

For the price of one drum seeder, you can buy two needle seeders, set up parallel seeding lines, and get nearly the same quantity of trays sown per week with a lot more flexibility.

What are large plug growers using for seeders? Jenny Kuhn, seeding supervisor at C. Raker & Sons, Litchfield, Michigan, uses two Blackmore cylinder seeders to sow 99% of all Raker's plug trays, even odd-shaped seeds. They run the trays slower and may have someone help move a few seeds from one cell to another, but they see no need to use a needle seeder. Tagawa West Coast Growers, Arroyo Grande, California, uses a Bouldin & Lawson drum, Blackmore cylinder, and KW needle seeders to sow four different tray sizes and a wide product listing. Suncoast Greenhouses, Seffner, Florida, has a Blackmore cylinder, turbo (vacuum-tip), and needle seeders to handle seeding needs. Headstart Nursery, Gilroy, California, uses one Hamilton cylinder and one KW needle seeders to sow a wide range of cut flower seeds. Knox Nursery, Winter Garden, Florida, uses drum seeders from Blackmore and Hamilton.

Designing Your Setups

No matter what your current seeding equipment setup is, if you're thinking of getting new or more equipment, you need to ask yourself some hard questions first. When visiting trade shows, talking to suppliers and other growers, and reviewing product literature, keep the following questions in mind:

1. How much space do you have for flat filling, seeding, covering, and watering in plug trays? Can you move some equipment closer together or put them in line, connecting them with conveyors?
2. What seeding capacity do you need? Your capacity is based on how many plug trays you need to sow each week, especially in your peak sowing weeks. Many plug growers buy a high-volume seeder that can handle 1,200 trays per hour, but then never come close to using this capacity, as the average sowing run is ten trays or less per variety. That's like buying a Ferrari, but never being able to drive it over 45 mph.
3. How many different tray sizes are you using? Is the seeder flexible in handling different tray sizes?
4. How many seeds per cell are you sowing for each crop?
5. How well does the seeder handle odd-shaped seeds?
6. What's your budget? Seeding equipment isn't cheap, so make sure you know how much you are willing to spend, and what the payback period is.

7. When it comes to retrofitting, will your current equipment work with the new equipment? Can equipment be added later?
8. What kind of training and service does the manufacturer offer?
9. How much are you willing to compromise between cost, speed, and reality?

The more realistically you answer these questions, the better your equipment purchase will be. Only you know how much space you have, how many trays you need to sow, and who is going to sow them. Remember, the seeder and seeder operator are not always at fault for poor seed placement or germination. The flat filler, dibbler, topcoater, and watering tunnel greatly influence seeding accuracy, amount of covering, and how much water is available for germination. Make sure to consider more than just the seeder when you decide to improve your seeding area.

The Ultimate "Dream" Seeding Room
Larger plug growers need at least two to three seeding lines to handle all types of seed, tray sizes, and tray quantities during the season. This gives them the flexibility needed to offer a wide range of varieties in a wide range of tray sizes. The type of seeders they use are a matter of personal preference and need. While some plug growers can get all the flexibility they need from drum seeders, others prefer using both drum and needle seeders.

The more volume that goes through the seeding line, the more desirable it is to integrate all the equipment. In-line seeding means everything is tied together, from flat filling and labeling to the watering tunnel and putting the trays on a bench. Conveyors connect the different pieces of equipment, and electronic eyes control the flow of trays.

Knox Nursery set up two integrated seeding lines at their new facility. The lines include a tray destacker, tray filler, dibbler, seeder, topcoater, watering tunnel, and a robot for loading trays onto a containerized benching system. Tray labeling and bar coding will eventually be done automatically, as well.

Most of C. Raker & Sons' system is in-line, although flat filling is still done separately. They load trays manually onto the seeding line, and take them off manually at the end of the conveyor to place on the bench.

Tagawa West Coast Growers has one flat filler serving two integrated seeding lines but is planning to have a flat filler for each seeding line and a robot for placing trays onto the benches. Labels are computer printed but are still placed on trays by hand.

Cutting-Edge Plug Technology

Some of the newest technology coming into the seeding area is the use of computer-generated labels automatically applied to trays. Bar codes and a scanner allow for identification and tracking throughout the greenhouse. Several manufacturers are working on bar-code label equipment for plug trays. Jenny Kuhn at C. Raker & Sons says they're planning to install bar coding and scanners, along with a computer terminal, at their seeding lines. The seeder operator can then call up on the computer the variety required, and the correct number of plug trays will automatically be filled, labeled, bar coded, and sown. The trays can be tracked in the greenhouse with portable scanners, allowing workers to quickly verify and update each tray's status.

A continuous plug tray system using injection-molded plug trays with no lip on the edges is now being tested. These reusable plug trays are designed for large growers using robotic transplanters. Because they have no lip on the edges, plug trays run continuously through the transplanters with no wasted space or machine motion. The trays are heavier in construction and can stand up to the wear and tear of robotic transplanting.

Finally, a new robotic replugging system that will provide 100% usable plugs is being trialed in the United States through Flier USA. This system can be fully or semi-automatic. It uses a computer vision system to scan each plug tray and determine empty or non-usable seedlings. These plugs are then dislodged from the tray by air ejection, and usable seedlings are replugged into the tray. With the increasing use of robotic transplanters, there's increasing demand for completely full plug trays. For large plug growers who sell to other growers, replugging trays to acceptable guaranteed stands can be a big cost. Robotics can make the process quicker and cheaper.

The ultimate "dream" seeding room would be clean and quiet, with controlled environmental conditions to help these expensive and complex seeders work better. The equipment would be designed to handle all steps in the seeding process, from filling and labeling trays to watering and placing on benches. The various pieces of equipment would be integrated to form an in-line system. More than one seeding line would be available to handle large volumes or multiple tray sizes. Bar coding would help the grower track trays throughout the greenhouse. And, of course, all of the equipment would work properly all the time and be cost-effective for labor savings, accuracy, and flexibility.

Summer 1997.

Is It Finally Time to Buy a Transplanter?

by P. Allen Hammer

Automatic transplanters have invaded bedding plant greenhouses. Today there are many different transplanters to choose from, ranging in price from $23,000 to more than $115,000, with transplant rates varying between 3,600 to 50,400 plugs per hour (100 to 900 flats per hour). So many growers without this technology are asking, "Is it finally time to buy a transplanter?" I certainly can't answer that question by myself, so I have contacted transplanter manufacturers, growers with transplanters, and growers not using transplanters and presented all of them with a list of questions. The response has been wonderful and a real education for me. This is a summary of what I have learned, and hopefully it will help growers who are wondering whether they should take this next step in automation.

No Simple Economics

Many growers would like to make a transplanter purchase based on very simple economics. If it cost twenty-five cents to hand transplant a flat, then an $80,000 transplanter would be paid off after transplanting 320,000 flats. Unfortunately the question isn't this simple to answer. There are many other factors to consider in a transplanter purchase. Responses from growers already owning transplanters provide a great deal of insight.

Interestingly, all the growers I have surveyed are satisfied with the transplanter they purchased, agree it saved production costs, say the transplanter works at the stated rate, and state that they would still decide to purchase a transplanter after their experience with the technology. The growers expect a payoff period for the transplanter of two, three, five, or six years, with most stating a three- or four-year payoff. Major advantages of automatic transplanters are labor savings, transplanting speed, production quality, uniformity of transplanting, and ability to turn flats during the spring crunch time. Other benefits include consistency of planting, seven-day work weeks, and increased flat production in the same greenhouse space. Many growers give transplanters very high marks for improved production quality from uniform and consistent transplanting.

Nothing Is Perfect

The growers also describe the major problems they've experienced in using transplanters. There's a learning curve to using automatic transplanters, they say. You can't have a transplanter delivered, plug it in, and start to transplant. There's downtime with transplanters that must be accepted. Transplanters require constant attention and minor readjustments. Often the same crew used for hand transplanting doesn't work well with automatic transplanters, and many growers report difficulty in finding quality employees to run the transplanter. The problems of different plug flats (sizes and shapes), inconsistent plug flats, and the shutdown required to change configuration for cultivars are also listed as problems. And certainly the cost (initial investment) of available transplanters is a problem for many growers.

Growers state that changing from hand transplanting to machine transplanting often requires production changes. The plug flat quality becomes a major issue. Poor plug flats can't be used with automatic transplanters. All available transplanters also depend on a good root ball to pick up and move the seedling. Most of the growers have also opted to use specific plug flats and finish flats as much as possible to reduce readjustment requirements. Some growers have also changed plug and flat sizes to better utilize transplanters. In general, machines are less flexible than people are.

Advice from Growers

I have asked growers what advice they would give to someone considering a transplanter purchase. Their responses are very thoughtful and helpful.

Don't consider labor costs alone in a transplanter purchase. Nearly every grower says transplanting quality, which can't be attached to the cost of transplanting, has been improved with the machine. Ask yourself if you really need a transplanter and whether you can justify the payback period in your specific greenhouse operation. At the same time you must look ahead when deciding to purchase a transplanter. Will a transplanter improve your bedding plant production? Do you have or can you afford to hire a staff capable of handling an automatic transplanter? Look at the versatility of a specific transplanter, and select the one that best suits your needs. There's simply not one machine that is best for all greenhouses. And lastly, as one proud transplanter owner states, "My advice is just do it [buy a transplanter]."

Among the most important factors to consider in a transplanter purchase, growers list high labor cost, lack of available labor, and the need for the machine's speed and accuracy to improve productivity. Those factors that influence growers who defer a decision to purchase are cost, long payback, lack of flexibility in flat size, no additional uses of the transplanter other than bedding plants, and the small numbers of flats produced by some growers.

Details to Consider

As a part of answering the original question, I have spent time reading transplanter literature and viewing videotapes of the various transplanters in action. Much information is available on the different transplanters, and it is not my intention to rewrite or even summarize that material. But here are some things I think you should consider and ponder as you review the available transplanters: A medium-sized grower produces around 60,000 flats/year while a high-volume grower produces 100,000 or more flats/year, as best I can determine from the positioning of the transplanter manufacturers. Different machines are available to suit various-sized greenhouses.

Carefully consider the transplanter's flexibility. We're at the very edge of machines that will transplant flats, hanging baskets, and pots. Also consider planting quality, uniform planting depth, faster finish, and more turns in the greenhouse as very important issues when considering the purchase of a transplanter. It's often very difficult to attach a real dollar figure to how these issues affect your return in increased profits.

The grower and greenhouse staff must be committed to using the transplanter. Patience and a desire to make the machine work are required. The machine operator must have the technical ability to operate a fairly complex machine and be trained on transplanter use. When reading transplanter literature, watch for special requirements that include an air compressor, 208/240 VAC single-phase 20 amps, or even a computer. These are sometimes listed as extras.

Estimates of planned downtime are often given in the literature. For example, changing plug flat size can require one hour, while changing finished flat size requires fifteen minutes, and changing cultivars requires a stop and restart. This does vary somewhat between machines, however. And certainly the purchase of such a complex machine requires outstanding sales, service, installation, and available space for parts. Some manufacturers have a twenty-four-hour technical support team that can be extremely valuable when you have a problem that needs a quick answer.

Automatic transplanters are a part of our bedding plant industry. I would strongly suggest you evaluate this technology. There's a transplanter available to fit the needs of many diverse greenhouse operations. They are not limited to just the high-volume grower.

Another important concept that has come from my discussions with growers is prefinished flat production. Some growers are working with the idea of justifying transplanter cost by custom planting for other growers in their local area. This could certainly be a win-win situation for both the transplanter owner and the small greenhouse.

I'm not sure it's time for every greenhouse to purchase a transplanter, but I am sure it's time for every greenhouse grower to take a look at what is available and decide whether it is a good investment for his own greenhouse operation. Each grower must individually answer that question.

Summer 1996.

Grasping the Less-than-Obvious

by Chris Beytes

Repluggers have been around since about 1994, and today at least three manufacturers, Visser, Flier, and Cherry Creek Systems, build them. Flier and Cherry Creek are both pushing hard with this new technology, but Visser is still the leader in plug grading, with machines in use for at least four years at plug ranges throughout Europe.

I can finally report that there's one on this side of the Atlantic, at C. Raker and Sons, Litchfield, Michigan. When Raker's makes an investment like this, it's worth finding out why.

A Vision of the Future

Visser calls its machine the Tray Vision 100, made up of a TIS (Tray Inspection System—the vision/blow-out portion of the setup) and TFS (Tray Filling System—the replugger). Raker's system uses one TIS and two TFSs because the TIS can grade and blow out about 500 trays per hour, while each TFS unit can handle about two trays per minute—about 120 trays per hour. One TIS can easily feed five or six TFS units.

About ten to fifteen days after sowing, plugs are brought into the grading area and sent through the TIS. Two video cameras take pictures of the tray. Software, with parameters set by the operator, determines which cells are duds.

After having its picture taken, the tray moves to the blow-out unit. Grippers grab the tray and tip it down over air nozzles. The computer directs the nozzle to blow out the dud plugs. The tray swings back up, then moves by conveyor to one of the TFS units for patching.

The patching process works like an automatic transplanter: The first tray of any batch is used as a source of replugs. (The operator tries not to send a nice tray through the system first for this reason.) Trays come into the machine, and another camera tells the TFS where to pick up plugs from the source tray and where to put them into the tray getting patched. It's fascinating to watch the gripper moving rhythmically back and forth between the source and the patch tray, replacing about one plug per second. Once filled, another operator double-checks the finished tray and sends it back into the greenhouse.

The Returns

Payback for the $500,000 system should be about three years, says Gerry Raker, co-owner. Some of that comes from two obvious sources: labor and space—Raker's has gone from twenty employees working a manual patch line to five employees running the automated line. The other fifteen employees still work in the greenhouse doing manual patching and other tasks. And instead of grading just before shipping, Raker's now grades just after sowing. This frees up greenhouse space that used to be devoted to empty plug cells. This can be a major gain: If you grow plugs that average 90% usable plugs and patch them to be 100%, that's a 10% gain in space—one acre in a ten-acre range!

But for Gerry, the greatest advantage may be in an area that others might overlook—improved customer service. Because Raker's is now grading at least two weeks earlier in the production cycle, they know what their availability is much earlier and can pass that information along to their brokers. While it's a challenge to take advantage of this, it's a major benefit to Raker's, their distributors, and the end customers.

Beyond the Obvious

Gerry's ability to think outside the lines is part of what has made Raker's a successful business, and I had little doubt that Gerry's reasons for investing in a replugger would go well beyond the obvious benefits.

If you're considering a similar high-end investment, take a lesson from this. Look beyond the obvious, and seek out the intangible or hidden gains that others may not take advantage of. If every plug grower with a replugger saves on labor and space, there's no competitive advantage to owning one. The same is true for transplanters, Dutch trays, watering booms, or carts.

However, if you continually pursue the benefits that others miss, you'll always have an edge.

Acres & Acres, November 1998.

Sowing Seeds of Change
by Bill Sheldon

"We're going through an Industrial Revolution in horticulture right now. This is our biggest year for new products," is the easy response from Robert Lando, president of Oberlin, Ohio-based Flier USA when asked about automatic seeder technology. "Small and medium growers can have an integrated [seeding] line for less than $40,000."

McMinnville, Tennessee's Bouldin & Lawson is also on the cusp of change. Engineer Graham Goodenough hints at that, saying, "There's a good deal of secrecy this time of year. We're working on several new machines. We'll have revolutionary aspects and some surprises."

David Steiner, vice president of Blackmore Company, Belleville, Michigan, a pioneer in automated seeders, is equally succinct: "What growers have now is a plant factory. And this means the smaller grower is getting more flexibility at lower cost."

Seed Technology: People Are Key

The focus seems to be on technology, but people remain the key. Growers are absorbing expansion with equipment, not more personnel. Or they're seeding faster and more accurately to use people more wisely.

Two growers with different operations stress the people portion of the technology formula.

Tim Raker, an owner of C. Raker & Sons, Litchfield, Michigan, has seven acres of houses at his twenty-two-year-old plug production business. "The labor issue is tough," Tim says. "You have to be progressive, offer good wages and benefits. Find people who are interested in what they're doing. We have to teach people to manage the equipment. Don't try to replace people with equipment, just get more efficient."

Martin Stockton is head grower of two-year-old First Step Greenhouse in Orange County, California. He has a highly automated 70,000-square foot facility. "From the beginning we underestimated the key position of the seeder," Martin recalls. "The operator controls our profit and success. We have to dedicate time and effort to train and reward this person.

"There's an art and finesse to seeding," he explains. "We need a person we can trust more and pay more to get us uniform and consistent production. This is hard to pencil. What's uniformity worth to you?"

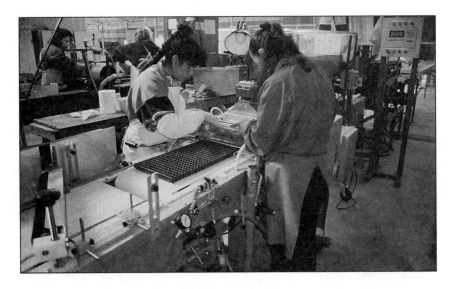

A good seed line starts with good people, such as those employed by Ball Tagawa in California. Only you can make the most of your seeder, whether needle (shown here), drum (in the background), or cylinder.

One of the country's largest growers also points to personnel, but with another twist. Sim McMurray is the head grower at Metrolina Greenhouses, Huntersville, North Carolina, a few miles north of Charlotte. This year Metrolina will have 68 acres of greenhouses and 300 peak-period workers, and will grow 100 million plugs, all for their own use.

"Finding help is the challenge of the industry. We've had a problem with that," Sim admits. But he adds that personnel needs expand into maintaining the line. "You almost need a factory rep on site," he says. "If a production line goes down, you'd better have quality maintenance people there to minimize your down time."

Roger Styer, an internationally known plug specialist, consultant, and author of *Plug & Transplant Production* sees the people/technology conundrum from two perspectives: "Growers make the mistake of not having good enough maintenance or a trained operator," he says. "Good equipment and poor training hurts."

Roger says that where the seeder is located can be as important as what kind of seeder you use. "[Growers] also don't seed in as protected an environment as they ought to. Cold, wet, hot might mean machines don't run as well." This is also important for the operators, who need more creature comforts.

"More education on equipment is needed. There's not as much of an off-season anymore, and machines or parts can't be sent back to the distributors as conveniently. This means service has to go to the machines." Roger puts some of the burden on manufacturers. "Seeder companies need to do better. There is too much limping along; they need to provide better help."

He adds that he thinks growers are ready to pay for necessary maintenance and training, and that they're looking for better, faster, more flexible machines. "The growers are pushing a lot of technology back on the distributor and manufacturer. There's also a lot of pressure coming from Europe. There's need for more accuracy, and for machines to handle a wider range of seeds."

Cylinders Roll On

"Speed is not as important as efficiency," says Dr. Dave Koranski, possibly the plug industry's best-known spokesman and champion. "Labor costs sometimes make 'faster' important. But the critical issue is the number of useable seedlings. The proper seeding operation can make the difference between 95 and 99% useable seedlings, and that can mean a lot of money. This is a bigger issue than anybody in North America understands." Dave explains that the issue is not just about equipment, "but instituting processes and procedures to do seeding right. You want consistent operations hour by hour."

Which gets Dave to the grower's decision-making process in buying a seeder. "Analyze the payback for your own operation, not for something that doesn't fit your greenhouse. Decisions should be based on your [business] and your costs, not your emotions. You have to look at the short- and long-term paybacks. Travel. Ask questions. Be critical. Do your homework."

While precision is an essential element of a seeder, growers are definitely looking for speed and efficiency in their sowing lines. The answer seems more and more to come from cylinder seeders, which combine the versatility of needle seeders with the speed of larger drum seeders.

Scott Swift has 100,000 square feet of plug production at Swift Greenhouses, his family's Gilman, Iowa, business. They grow liners of 600 varieties of perennials and herbs. Scott has moved from a turbo needle seeder to a cylinder seeder.

"We're phasing out the turbo because we have several people who can run the cylinder," Scott explains. "They can do in part of a day what it used to take all day on the turbo. This frees them up for other tasks."

Martin Stockton's California operation has grown 70% in its two years, and he, too, has moved from needle to cylinder seeders. "We should have bought the cylinder in the beginning," he says of his decision. "We didn't because we were relating equipment to what we were doing at the time, and we were worried about spending twice as much. But with the cylinder we're not looking at payback. We were at sixty-hour weeks, but with the cylinder we get back time and get better seed placement, which reduces our error percentage—a vital, key situation."

Tim Raker has also made the transition over to cylinder seeders. "We've had drums and needles and went exclusively to cylinders a little more than five years ago. We wanted versatility and easy changeover. We often do four different jobs on the same cylinder head. We're into our second generation of cylinders. Now we have three Blackmore cylinders that do all our seeding."

Roger Styer says this isn't uncommon. For the plug growers he's visited, "Blackmore cylinders have become the high-volume seeder of choice."

Manufacturers, such as Blackmore's David Steiner, spend a lot of time evaluating how their customers can get the most from their seeders, whether needle, cylinder, drum, or simple plate seeders. David suggests that the best seeder is the one that lets you spend the least time using it.

"Especially for the smaller grower, time is getting more valuable. Why have your best people standing around all week on a seeder? Have them seed two days and actually grow plants most of the time. That's where grower value is."

Another consideration is greenhouse space use. "The faster equipment also makes better use of greenhouse space," David says. "When you're heating to 75°F in January to seed only a couple of days a week, you're saving money."

While these growers are now using cylinder seeders for virtually all of their production, Robert Lando of Flier USA is more cautious. "We make drums and cylinders and sell a lot of them. But they each do not necessarily work on every seed," he says. "Growers teach me that each machine has a reason. Large growers, especially, use each machine that's better for certain things. The grower decides which way to go by looking at the job as just one corner of the day, not just what has to be sown. Varieties change, how many seed sizes are involved, what short runs do you have. They want to reduce changeover time.

"Seeding alone might be only 20% of the total production time," Robert concludes. "Make sure the seeder can perform all the tasks you need."

Graham Goodenough of Bouldin & Lawson urges growers not to invest in excess capacity, and to increase speed only as their business increases.

"A typical three- to five-acre grower can build his business properly with one fully automated precision needle seeder for about $16,800," Graham says. "Distributors shouldn't quote silly production numbers. Six hundred flats an hour is enough for most people. Accuracy is the key. It's of no value to have big speed when you aren't going to or can't use it."

Graham adds that there are lots of changes coming on needle and drum seeders. "We're increasing the range of our seeders. I think we all may see something radical in a couple of years."

David Steiner says Blackmore sells roughly an equal number of cylinder and needle seeders. "But this includes selling a lot of used seeders from the '80s that are available. We probably sell two-to-one used to new turbo seeders. If you look at new cylinders compared to new needles, it's about three- or four-to-one."

For a fully automated, in-line system like Metrolina's, they use cylinder seeders for everything except marigolds, zinnias, daisies, and some perennials. Sim McMurray estimates these products are about 5% of their total workload. Sim also says he expects equipment costs to come down in a few years, opening opportunities for smaller growers.

These growers, distributors, and consultants clearly believe in what they're doing and have opinions that conflict with each other's experiences, which probably reinforces Dr. Dave Koranski's comments on growers being careful to look at the seeding process for their own needs.

And, then there's one other highly scientific analysis suggested by Flier USA's Robert Lando: "Maybe it's just because growers like their toys?"

Summer 1999.

The Transplanter/Tray Relationship

by Chris Beytes

You say low-buck transplanters have piqued your interest, and you're ready to invest $30,000 to get set up for automatic transplanting? Great! However, before you enter the magical world of transplanters, you have to consider one very important thing: your plug tray. When transplanting by hand, you can pull any kind of plug from any size of plug tray—your fingers are infinitely adaptable. It doesn't matter if you buy plugs from six different suppliers in six different shapes and sizes of trays.

Buy a transplanter, and suddenly that planting freedom comes at a price: You can choose a high-end machine costing more than $100,000 that adapts to a variety of plug trays, or you can buy one of the lower priced machines, then invest thousands more to adapt it to the specific plug trays you like to use.

Whether you grow your own plugs or buy them in, you have to make some adjustments to your way of looking at plugs before you can start transplanting them automatically.

Tray Size Matters

The first plug tray challenge you encounter is the incredible range of shapes and sizes that are on the market—anywhere from 50 to 800 plug cells in a typical eleven-by-two-inch tray; in square, round, or multi-sided shapes; with straight or pyramid-shaped sidewalls; deep or shallow cells; large or small drain holes . . . the combinations are endless.

However, thanks to growers who've pioneered automatic transplanters, the industry is beginning to standardize to several primary plug sizes: the 512 and the 288. The 512 tray is less expensive, and you can squeeze more of it into a greenhouse than you can a 288, but it takes longer to finish. The larger 288 plug is said to be the most efficient to grow, provided you have the propagation space available. It also finishes more quickly, especially in larger containers such as four-inch pots or 1801 packs. But it's more expensive to grow or to buy in. Another size that's growing in popularity is the 384, which transplants well into 606 jumbo packs.

One part of this plug size standardization comes from transplanter efficiency. For example, transplanters with sixteen or thirty-two grippers work most efficiently when they can pull from a 512 tray, which is thirty-two plugs long. A 288 tray is twenty-four plugs long, efficient for a machine with twelve or twenty-four grippers. Other combinations will either leave a few rows of plugs in the tray or leave grippers empty during a transplant cycle.

The other part of standardization comes from the differences between trays of the same size: All 512s aren't created equal. While some manufacturers have trays that are essentially interchangeable, others vary a bit more in design, such as drain hole shape or size. Transplanters, which use various mechanical means of dislodging plugs, often handle trays from one manufacturer better than those from others because of these differences. Some transplanters are more flexible than others. For instance, they may not use ejector pins, or they may have pins that are somewhat flexible and can adjust themselves to tray diversity.

Tray manufacturers—under pressure from plug growers who are under pressure from transplanter owners—are beginning to accept the idea of standard plug tray designs that will work in the commonly used transplanters. At least two plastic companies now have trays that are essentially interchangeable, and a third has a similar design. Transplanter manufacturers are also beginning to recognize the need to build machines that will handle the commonly used trays, rather than designing machines around unique tray configurations.

What's the Answer?

One plug tray manufacturer offers these three suggestions for anyone getting ready to buy a transplanter:

First, if you're currently growing or buying plugs in something other than 512s or 288s, and you're considering an automatic transplanter, change plug trays now—other sizes may not be as readily available down the road. Also, your transplanter choices will be much wider, and you won't have to spend money customizing your new machine or changing over when your favorite oddball plug tray is no longer available.

Second, if you're growing your own plugs, choose a reputable tray manufacturer, and stick with them. Get to know their product, its quality, and its consistency. Be loyal to them so they'll be loyal to you.

Third, if you're buying plugs in, make sure your suppliers know you're about to become mechanized. Find out what trays they're using, and make sure they'll be compatible with your new machine. Communicate with them regularly, and make sure they're providing top-notch plugs with good root systems.

You may also want to adjust your shipping schedules: One large shipment for a two-week transplant window may have been fine when planting by hand, but a machine is more particular. Get two shipments one week apart instead so plugs won't overgrow.

You'll have plenty of challenges fitting a new transplanter into your production regime; doing your homework up front on your plug trays could mean one less headache.

Hardware, October 1998.

The CE Tray: Savior or Threat?

by Chris Beytes

The idea of uniform container standards has been argued for years, and the topic has become especially critical with the advent of automated plug sowing and transplanting equipment.

Finally, the industry has its first container standards, which came out of a meeting initiated by Flier USA President Robert Lando. (Flier manufactures seeders, transplanters, tray dispensers, and other container-handling equipment.) Back in August, Robert invited the major plug tray manufacturers to a meeting to discuss the idea of creating standard dimensions for plug trays.

"We didn't think it was our place [to set the standard]," Robert explains. "We just thought that if this meeting could be the vehicle to get there, then that would be great."

The group had no problem agreeing that the industry would benefit from common elements, and that it wouldn't be difficult to agree on the elements needed to be standardized: cell spacing, tray width, drain hole size and placement, height, and minimum tolerances of each dimension. These standards would still allow enough design flexibility so that each manufacturer could still create unique products.

"It's the only thing that makes sense," says David Steiner, vice president for Blackmore Company. Blackmore not only is a leader in trays and seeders, they also distribute Harrison and Rapid transplanters, so they understand the mechanical benefits of a plug tray standard. "When you have a $500 dislodger it's not that big a deal," says David. "But when you have a $100,000 transplanting machine that you can't use because somebody didn't make the right kind of trays, then it becomes an issue. That's what has forced the standardization in the industry."

Jeff Kissenger, marketing manager for Landmark Plastics, agrees with David. "We view it as a benefit to the industry." Jeff, however, wouldn't reveal when they'll begin manufacturing and promoting CE (Common Element) compatible trays. Other manufacturers say they'll have CE trays available this spring or as customers begin to ask for them.

Like Flier, Bouldin & Lawson, which builds transplanters, sees the standard as "an excellent idea." Says Marketing Manager Jim Fowler: "After we adapt to [the new] dimensions, it will actually make our jobs as manufacturers a little

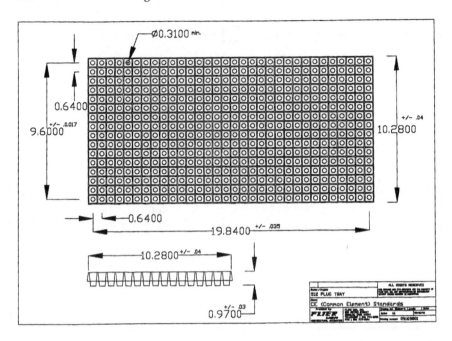

Common elements of the CE tray, such as the center-to-center dimensions of the cells, make it the first true standard for the industry.
Illustration: Flier USA

bit easier because we won't have such a variety of setups to do. We won't have to custom build every machine."

Possibly the biggest benefit of the standard is for growers buying in plugs and using an automatic transplanter. Right now, you have to make sure any supplier you use is growing in a tray that's compatible with your machine, or else you have to spend $800 to $5,000 or more adapting your transplanter to a different tray. With the CE tray, you can configure your transplanter to that standard (if it isn't already). Then, any CE tray you buy from any source will run through your machine. Plug growers we've talked to say they haven't yet started discussing the trays with their customers, but they think a standard is long overdue.

"I think everybody would be better off to conform to a standard configuration that still allows flexibility and interchangeability between plug sources," says David Wadsworth, owner of Suncoast Greenhouses.

Speedling's Barry Ruda sees the new tray as a marketing tool: "There are some definite advantages to going to [a standard], because it's a real problem to sell to new customers because of transplanters [and their need for specific trays]."

What's the Downside?

Changing over to CE trays could require some more investment in your existing equipment. You'll have to make sure your existing plug popper or transplanter fits the new tray, or else have it modified. If you're buying a new machine, just specify that it be set up for CE trays, and you'll be able to transplant from any source without any additional costs. If you grow your own plugs, you don't have to do anything: Tray companies assure customers that they'll continue to make their existing trays as long as there's demand for them.

A bigger issue than costs is that standards are fine—as long as we're the ones setting them. What if chains decide on a standard pack, for instance? Some in the industry express that fear. However, there's never been any evidence that chains are interested in those kinds of standards. If anything, we have too many containers because of the chains' interest in always being different from their competition.

In fact, standards for finished packs and pots would go a long way toward reducing equipment costs and increasing innovation. The CE meeting came about through an earlier meeting between Flier and two hanging basket companies. The topic: Could they agree on standard rim and hanger designs that would let Flier design a machine for automatically attaching basket hangers?

However, Robert emphasizes that the CE movement has no intention of impacting finished plant containers, except to make growers' lives easier.

"There's always a place for variety in the finished plant marketplace," he says.

Hardware, February 1999.

Chapter 5
Crop-by-Crop Guide

Key Points for Plugs

by Dr. Roger C. Styer

Much has been written and said about plug growing in the past few years, and with all of this information out there, you'd think plug growers would have no problems growing any crop from seed. However, I still hear from growers who have difficulties with one crop or another, and often they have the same problems in common.

Here, I've summarized briefly the key points for success with twenty-four popular bedding crops. These points cover the areas I think every plug grower must know to have success with these crops. Of course, there are many factors that influence the quality of your plug crops, such as water and media quality, nutrition, the greenhouse environment, and others. But focusing on the following essential points for each of these crops will at least help you have a chance for consistent success.

Ageratum
- Avoid high Na (sodium) and salts.
- Watch overdose of growth regulators.

Alyssum
- Multiple-sow, but avoid too high a density or you'll have damping-off.
- Keep lower leaves green.
- Watch for downy mildew.
- Maintain height control with DIF or B-Nine.
- Avoid high light, which causes yellow leaves.

Begonia
- Keep uniformly moist until first true leaf is half expanded—begonia has shallow roots.
- Dry off media surface after this point to control algae.
- Avoid algal crust.

- Feed early when cotyledons are expanded, include some ammoniacal nitrogen (NH_4) for leaf expansion.
- Stunted plants are due to drying off too soon or underfeeding.
- Tone with moisture stress, high light, cool temperature, or high-calcium fertilizer.
- Avoid watering in the afternoon on sunny days which causes burned leaves.
- Control fungus gnats and shore flies.

Celosia
- For best germination, double-seed, keep uniformly moist, and watch for high media EC.
- Need early growth regulator applications for tall varieties.
- Avoid stress to plugs, or they will bud or bloom in plug trays.

Coleus
- Slower colors need more feed.
- Use high light or growth regulators to control height.
- Avoid rhizoctonia by controlling moisture.

Dahlia
- Difficult to sow.
- Use growth regulators correctly to control growth, and keep plants fed.

Dahlia

Dianthus
- Avoid high salts and letting them dry out too much.
- Control growth on vigorous varieties with proper growth regulator applications.
- Purple leaves may be a sign of high light, cool temperatures, or phosphorus deficiency.

Dusty miller
- Double-seed.
- Avoid high temperature and high light, or temperatures that are too cool during germination and seedling growth.
- Control alternaria leafspot by moisture management, air movement, and fungicides.
- Avoid Bonzi drift.
- Stress plants to produce dusty leaves.

Geraniums
- Keep media pH between 6.0 and 6.5 to avoid micronutrient toxicities.
- Control height with growth regulators, but watch out for overdosing with Bonzi.
- Watch for pythium and botrytis.
- Keep crop growing to get buds on time for ten-week geranium crop.

Gerbera
- Keep on the dry side for better germination and growth to avoid stunting.
- Use high light and high feed levels.
- Control thrips, leafminers, and aphids.
- Control powdery mildew and leafspots.

Hypoestes
- Avoid high salts.
- Keep under some shade to prevent leaf curl and bleaching.
- May not need growth regulators.

Impatiens
- Uniform moisture during germination.
- Dry down after cotyledons come out for best rooting, and control early stretch.
- Use primarily high calcium feed, but avoid overfeeding.

- Keep media pH between 6.0 and 6.5.
- Use growth regulators as needed, but avoid stunting plants.
- Use moisture stress, high light, and high calcium feed to initiate branching and budding.

Lisianthus

- Dry off media surface after germination is finished—lisianthus has deep roots.
- Control algae.
- Keep well fed.
- Avoid temperature above 78°F in first four weeks to avoid rosetting.
- Keep media pH between 6.0 and 6.5.
- Control pythium, fusarium, and botrytis.
- Control fungus gnats and shore flies.

Lobelia

- Use multi-seed pellet.
- Adjust growth regulators by vigor of varieties.
- Control damping-off and high salts.

Marigolds

- Hard to sow.
- Cover after seeding, but make sure you can see ribs of tray after watering.
- African marigolds like short days to flower and may need more growth regulators later in season.
- Avoid stressing French marigolds to keep from budding and flowering in plug trays.
- Adjust growth regulators by variety.

Pansy and viola

- Use primed or pregerminated seed when germinating in warmer temperatures.
- Cover and keep uniformly moist until seedlings first start to hook up, then dry back to control early stretch.
- Keep media pH less than 6.5.
- Avoid overfeeding, and use a fertilizer higher in calcium.
- Practice good moisture management for good roots and root hairs.
- Use B-Nine, B-Nine/Cycocel tank mix, or A-Rest to control height.

Petunia
- Use pelleted seed.
- Dry down after cotyledons expand to keep seedling growth uniform—petunia has deep roots.
- Keep crop fed and media pH below 6.5.
- Control growth with growth regulators and moisture management, and adjust by variety.
- Use high light or HID lights in the North to provide long days for earlier flowering.

Portulaca
- Multiple-seed.
- Keep warm and on the dry side after germination.
- Adjust growth regulators later in season.
- Avoid damping-off disease.

Salvia splendens

Salvia splendens
- Keep on medium-dry side, if covered, for best germination.
- Use early growth regulators to control stretch.
- Avoid overfeeding early, but provide heavier feed later in crop.

Salvia farinacea *and species*
- Double-seed.
- *S. farinacea* need longer crop time than *S. splendens* types.
- Use growth regulators later in crop cycle.
- Keep temperatures warmer.
- Avoid high salts on foliage to prevent tip burn.

Snapdragons
- Dry down after germination is finished to control early stretch and get better rooting.
- Adjust growth regulators by vigor of varieties.
- Don't overfeed, and use high calcium.
- Keep media pH below 6.5.
- Control pythium, fusarium, and downy mildew.

Verbena
- Keep on drier side during germination by keeping moisture in just the covering.
- Control damping-off and leafspots.
- Avoid overfeeding and high EC.
- Adjust growth regulators by vigor of varieties.

Vinca
- Keep uniformly moist with covering until seedlings start to hook up, then dry back for better rooting and germination.
- Keep warm during germination and growth.
- Avoid frequent watering, and use good wet/dry cycle.
- Keep media pH below 6.5, and avoid overfeeding. Use high calcium feed.
- Watch for root rots, rhizoctonia, botrytis, and aerial phytophthora.
- Don't use Bonzi—it causes spots on leaves—and avoid overdosing with growth regulators.

Zinnia
- Keep crop time short.
- Use early growth regulator applications for tall varieties.
- Control leaf spots.
- Double-seed angustifolia types, and use longer crop time.

September 1999.

Seed Geraniums the Brodbeck Way

by Bruce Brodbeck

Twelve years ago, we at Brodbeck Greenhouse couldn't help noticing all of the volunteer geranium plants growing on the greenhouse floor around our plug seeder. Seeds that bounced from the seeder or plug trays during seeding seemed to be thriving on the ground floor without getting the care we gave to our official geranium crop.

Some of the problems we experienced with our geranium plugs included a tendency for the plant to damp-off late in the plug growth stage. When we transplanted these 288 plugs into the finished flats, there was also a problem getting the new transplants to take root. Every year we had more misses in the flats than we liked to see. Consequently, patching the misses was just part of our seed geranium production. Since then we've made a few changes to ensure our production is more efficient and cost effective.

Direct Sow for Flats

We originally sowed into 288 and 512 plug trays. During our initial trials of sowing geraniums directly into cell packs, we concentrated our efforts on achieving exact seed singulation. When seed costs nearly $0.05 apiece, any doubles quickly pushed aside any economies derived from this method.

We configured our seeder to place one seed in the center of each cell. The flat continued on to be covered with about ¼ inch of coarse vermiculite; then it was watered lightly to keep the seed and covering in place while it traveled to the greenhouse and was set on the ground. Once the flats were on the ground, they were given a more thorough watering.

In the past three years, we've changed our top coating to Scott's Metro Mix 360. It flows through our top coater hopper nicely, and we've found that its color change more closely reflects the moisture needs of the flat than vermiculite. With the flats now set down in one of our floor-heated houses, we keep them moist, usually watering them once per day. The heating coils in the floor are kept at about 100°F, which keeps the soil in the flat in the high 70s. The underground heat helps ensure that the germinating seed is receiving the necessary temperature, which may not be obtained solely with conventional overhead heating methods, as temperatures near the floor often vary from ambient temperatures. (We've experimented with covering these flats with a sheet of thin-mil poly sheeting, but the short germination time combined with low light levels at this time of year didn't seem to make much of a difference for the extra steps it took.)

Keeping the house at a relatively high humidity of at least 90% or more generates quite a bit of condensation on the inside layer of our poly houses. We installed a third inside layer of poly to collect any drips and channel them toward the eaves. Any drips lower the desired soil temperature, and consistent drips will erode the top coating on the flat and wash out the seed.

The seeds germinate in approximately five to six days, and over the next two weeks the likely germination percentage becomes fairly clear (usually fairly close to what's indicated on the package). To account for the expected loss in germination, we sow some "spare parts" in plug trays a few days after the direct sowing, usually in the amount expected not to germinate.

After three to four weeks or when the plants are in their fourth or fifth leaf stage, we move them into our growing house. We use a small portable conveyor to run the flats across. Then we fill in any missing cells or remove any misshapen seedlings with the ones grown in the plug trays. At this stage, many labor cost savings are realized. This process usually requires only ½ of the labor required for a traditional transplanting line, and with only one or two cells needing replacement per flat, production is nearly three times as fast. Flats are then loaded up and moved out into an area where the temperature is maintained at 75° to 80°F days and 60° to 65°F nights.

Media and Fertilization

The soil we use for geraniums is a 70% peat/30% yellow sand mix. The long growing time plus the rather close plant spacing in our 1203 packs, puts undesirable pressure on the plant to stretch. This mix helps carry the geranium from germination to shippable size.

For fertilization, we use a peat lite 20-10-20 blend as a constant feed at 50 to 75 ppm throughout the whole growing period. For growth regulation, we start applying Cycocel about three weeks after germination at ½ ounce per gallon of water. We make three to four applications approximately ten days to two weeks apart.

The method of application must include a few precautions to achieve desired results. We lower the pressure on the sprayer to about 50% of what's normally used for most applications. Then, being careful not to hit any of the leaves directly on the initial spray burst, we spray in a cascading manner, which lets droplets settle down on leaves.

Direct seeding naturally helps plants begin branching out at an earlier stage, which exposes more leaf surface and makes the Cycocel more effective. Before we refined this method, we were getting overdosed patches in the crop that disguised themselves, looking like viral damage with severe leaf burning.

We don't use any fungicides until late in the growth stage, when we treat with Chipco on a biweekly program along with other plant crops when prolonged periods of cloudy weather persist.

Direct Sow for Pots

We also direct sow into four-inch round pots in a fifteen-count shuttle tray. For this, we use a peat-lite soil mix containing very little sand. Because plants won't be grown as closely together as the packs, the lighter mix tends to hold more moisture and promote more side growth. We sow one seed per pot and set pots on heated floors to germinate.

After three to four weeks, we move plants off of the heated floors into another greenhouse for growing on. It's at this point that we pull out any pots that didn't germinate. After three more weeks or when leaves begin to cover the soil, we space pots to every other one in the shuttle tray.

We use the same growth regulation and fungicide program as we do for plants. Fertilization starts out the same way as well, and we increase to 100 ppm of 20-10-20 as the pots are separated and grown to maturity.

We use this crop to supplement our cutting geranium crop. Taking longer to come into color than our cuttings, the direct-sown crops usually fill late-season sales when cutting geraniums have already been depleted by earlier sales.

As plants mature, we generally see buds and blooms the first week of May. Ideally, our customers prefer them to be shipped with buds and three to four blooms per flat to show the color of the bloom and provide longer shelf life in the garden center. Our bedding plant shipping season begins around the last week of March, and with orders for geraniums, this early shipping gives us an opportunity to use the open floor space to space the remaining crop as it further matures. This extra spacing allows air to flow to the plants and deters the formation of yellow leaves under the top canopy that can take away from the overall display appearance.

For greenhouse operations that have year-round crop production, the long growing times and extra space required probably wouldn't make this method as feasible. We specialize only in the spring bedding plant season, shipping from the last week of March until the first week of June. Starting these geraniums in the last week of December fills an open time window in our plug production between slower growing crops, such as perennials, and the larger volume crops, such as our annual bedding plant plugs. Geraniums are the first crop to be grown in the greenhouses we use to germinate them in, and other crops aren't scheduled to be grown in those houses until after the geraniums have been moved out, so the growing space requirement really isn't an issue.

With the combination of good growing habits and cost savings, this direct sowing program has worked well for us over the past twelve years.

Culture Notes, December 1997.

Seed Geraniums

by Teresa Aimone

December is the month to start planning to sow your seed geraniums. Crop time for this popular bedding plant ranges from twelve to fifteen weeks, depending on growing conditions and variety.

Propagation

Sowing

After sowing into a sterile, well-drained plug mix, cover the seed completely with either a thin layer of medium (no deeper than the width of the seed) or #2 coarse grade vermiculite. Geranium seed don't require darkness to germinate, but they do require high humidity around the seed for

germination. Be careful not to bury the seed under too much covering or use a sowing medium that has large pieces of bark or other material that may cover the seed and hinder germination.

Temperature

Maintain a soil temperature of 72° to 77°F. Don't let the temperature fluctuate dramatically or rise above 86°F or below 68°F during the initial germination stage. These extreme temperatures result in poor germination and an uneven crop. After the cotyledons emerge, drop the temperature to 65° to 70°F.

Fertilization and pH

Seed geraniums prefer a pH of 5.8 to 6.5. This is higher than other bedding plants. The pH is critical for good seed geranium production. EC should be 0.75 mmhos. Once the cotyledons have emerged, begin feeding using a low ammonium fertilizer enhanced with calcium and magnesium. One fertilizer recommendation suggests a N:P:K:Ca:Mg ratio of 13-2-13-6-3, applied at 100 ppm.

Watering

Don't allow soil to dry out unevenly, as it can result in an uneven crop. Keep soil uniformly moist throughout the flat.

Lighting

Supplemental lighting can reduce crop time. Begin lighting approximately three weeks after sowing. Extend the daylength to eighteen hours per day, and continue for four to six weeks through Stage 2, Stage 3, and until plants are put into their final containers.

CO_2

Inject CO_2 at 800 to 1,000 ppm from sunrise until the vents open. Remember that both light and carbon dioxide are on the front end of the photosynthesis reaction. Both need to be used simultaneously to keep the equation in balance and produce the most uniform crop.

Culture Notes, December 1997.

Growing Lisianthus from Seed

by Brian Corr and Philip Katz

Most growers agree that lisianthus seedlings are the most difficult young plants to produce. Here are some tips for growing these challenging seedlings.

- Germinate between 72° and 75°F. Never exceed 82°F, as this will cause rosetting.
- Maintain near 100% relative humidity at the seed; use sufficient moisture to dissolve the seed pellet.
- Maintain medium to high levels of calcium and nitrate-form nitrogen in the media.
- Maintain substrate pH above 5.8.

- Remember the timing from sowing to transplant ranges from eight to twelve weeks. New varieties such as Ventura, Malibu, and Laguna are more vigorous in plug tray and require a shorter crop time.

Results of greenhouse trials in Southern California. Early season varieties sown week thirty-four, transplanted week forty-four; mid-season varieties sown week forty-six, transplanted to beds week four.

Lisianthus Variety Breakdown

Early season flowering			Mid-season flowering		
Variety name	Days to flower	Height (inches)	Variety name	Days to flower	Height (inches)
Ventura Deep Blue	246	37 to 41	Malibu Deep Blue	198	37 to 41
Heidi Deep Blue	258	35 to 41	Heidi Deep Blue	212	35 to 43
Ventura Blue	253	35 to 39	Malibu Purple	201	33 to 43
Ventura Blue Blush	255	37 to 41	Malibu Blue Rim	202	45 to 51
Heidi Pastel Blue	260	100 to 110	Heidi Blue Rim	206	39 to 43
Ventura Blue Rim	241	35 to 39	Malibu Rose	193	45 to 55
Heidi Blue Rim	255	35 to 39	Heidi Rose Pink	206	43 to 53
Ventura Rose	244	33 to 35	Malibu Lilac	204	43 to 47
Heidi Rose Pink	267	41 to 45	Heidi Lilac Rose	218	43 to 49
Ventura Deep Rose	277	33 to 39	Malibu White	193	49 to 51
Ventura White	244	31 to 33	Heidi White	200	37 to 47
Heidi White	248	31 to 35			

Culture Notes, July 1997.

Producing Fall Pansy Plugs

by Dr. Roger C. Styer

Here it is, the hottest time of the year and plug growers are trying to produce a crop that doesn't like the heat. Pansies are a cool-season crop, but the big market is now the fall season. Which means you need to overcome heat, humidity, rain, diseases, stretch, and other problems to produce a pansy or viola plug to transplant. In addition, the seed is expensive, making any mistakes in germination that much more costly.

To make the job easier and more productive, here are some recommendations by stages for producing fall pansy plugs. Also, specific problems that can occur with fall pansy plugs are outlined, including recipes to overcome or prevent them. Check your procedures against those listed to see where improvements can be made.

Stage 1: Sowing to Radicle Emergence

1. Cover uniformly with coarse vermiculite. Make sure to dibble trays properly to keep vermiculite in tray cells.
2. Keep soil temperature less than 80°F, if possible. This is more important with raw seed than with primed or pregerminated seed. Use primed or pregerminated seed whenever possible. Chamber germination should be done at 65°F.
3. Keep trays *uniformly moist.* Pansies need good moisture to germinate.
4. Light isn't needed for germination. Leave primed seed in the germination chamber for three days and raw seed in the chamber for four to five days at 65°F. Keep shade in the greenhouse closed all the time if germinating on benches.
5. Media pH should start at 5.5 to 5.8, with starter charge at less than 1.0 mmhos [using the saturated media extract (SME) test].

Stage 2: Radicle Emergence to Cotyledon Expansion

1. Keep soil temperature under 80°F, if possible. If radicles (initial roots) have emerged, pansies aren't as sensitive to higher temperatures. It's best to germinate in a cool chamber.
2. To reduce hypocotyl stretch (floppiness), reduce moisture levels to *medium amounts* once radicles come out and seedlings are hooking.
3. Keep shade closed except during early morning and late afternoon or early evening. Light levels should be 1,500 to 2,500 foot-candles.
4. Wait until the first true leaf is just visible (Stage 3) before feeding, as too much early feeding promotes floppiness.
5. No growth regulators are needed until Stage 3.

6. Media pH should be 5.8 to 6.0, definitely less than 6.5. Keep media EC less than 1.0 mmhos.

7. Apply a drench of Subdue (low end of rate) and Cleary's 3336 or Fungo (full labeled rate) after germination is finished to protect against root rots.

Stage 3: Growth of All True Leaves

1. Keep soil temperature under 90°F, if possible.

2. Use an alternating wet/dry moisture cycle to control stretch and promote good root development but *avoid wilting.*

3. Use higher light levels to keep plants shorter, but keep shade closed during the warmest times of day. Light levels should be 2,500 to 4,000 foot-candles, no higher.

4. Pansies are a low-feed crop, but you need to make sure that you expand the leaves and keep good leaf color without letting the petioles stretch. Use mainly 14-0-14 or 13-2-13 fertilizers. Use 20-10-20, 15-5-15, or 17-5-17 for the first feed and right after growth regulator applications for leaf expansion. Start with 50 to 100 ppm nitrogen after the first leaf is visible, then go to 100 to 150 ppm nitrogen. Feed every two to three waterings. You may need supplemental feeding with boron from Borax (½ ounce/100 gallon) or Solubor (¼ ounce/100 gallon) every two to three weeks.

5. Keep media pH at 5.8 to 6.0, definitely less than 6.5 to control thielaviopsis and avoid micronutrient deficiencies. Media EC should be less than 1.5 mmhos (SME).

6. Make the first application of growth regulators once the first true leaf is coming out on fast-growing varieties (see table). Use A-Rest at 10 to 12 ppm, and repeat in seven to ten days. You can also use a tank mix of 1,500 ppm Cycocel and 2,500 ppm B-Nine (one time only) if A-Rest isn't providing enough control. Reduce rates if growing in the North. For slower growing varieties, wait until the second true leaf is coming out to apply A-Rest, and repeat seven to ten days later, if needed. B-Nine at 5,000 ppm can be used during cooler weather or in the North on slower varieties.

7. Apply a drench of fungicides as described above every three to four weeks. Start spraying at least weekly with a rotation of Zyban, Daconil, and Chipco 26019 when foliage starts filling in to control leaf spots. Avoid keeping foliage wet during the night.

Stage 4: Toning and Holding before Transplanting

1. Keep soil temperature under 90°F, if possible.
2. Use alternating wet and dry moisture cycles to control stretch, but *avoid wilting.*
3. Keep shade closed during the warmest times of day, and maintain light levels at 2,500 to 4,000 foot-candles.
4. Feed only with 14-0-14 or 13-2-13 as needed, but keep media pH less than 6.5 and EC less than 1.2 mmhos (SME).
5. If you're shipping plugs, avoid applications of growth regulators within five to seven days of shipping. If you're using your own plugs, you can use growth regulators before transplanting to have control of early stretching after transplanting. However, wait ten to fourteen days after transplanting before applying growth regulators again.
6. Apply fungicide drench, as indicated previously, every three to four weeks. Keep up weekly sprays for leaf spots, and control moisture on leaves.
7. Avoid holding oversupply if plugs are in flower or disease is evident. Don't damage root hairs, as this crop is sensitive to root rots.

Problems

Thielaviopsis and other root rots

You need a good laboratory diagnosis to find out which disease you have. Keep media pH at 5.8 to 6.0 and EC less than 1.5 mmhos (SME). Avoid drying out or damaging roots. Apply fungicide drench every three to four weeks. Don't reuse plug trays. Reduce temperature and light stress on the crop.

Tip abortion, new leaf distortion, and stunting (boron deficiency)

Keep media pH at less than 6.5. Use supplemental feed of Borax or Solubor as described previously. Avoid overapplications of calcium, as this nutrient competes against boron.

Upper yellow leaves (high media pH or iron deficiency)

Keep media pH under 6.5. Avoid overapplications of 14-0-14 or 13-2-13 fertilizers, as they tend to raise media pH. Check your water alkalinity levels, and keep them at about 80 ppm with acid injection.

Lower purpling of leaves (phosphorus deficiency)

If plants don't have distinct spotting, then use fertilizers that have some phosphorus in them (13-2-13, 15-5-15, 17-5-17). High light levels and dry moisture will also cause phosphorus deficiency. If plants have distinct spotting, then check for leaf spots, and apply fungicides as described previously.

Fast-growing varieties	Slow-growing varieties
Accord series	Bingo series
Crown Golden	Crown series *(note exceptions)*
Crown Yellow Splash	Crystal Bowl series *(note exceptions)*
Crystal Bowl Blue Center	Fama series
Crystal Bowl Primrose	Happy Face series
Crystal Bowl Purple	Imperial series
Crystal Bowl Sky Blue	Joker series
Crystal Bowl Yellow	Majestic Giant Red & Yellow Bicolor
Delta series	Majestic Giant Rose
Majestic Giant series *(note exceptions)*	Majestic Giant Scarlet Shades
Rally series	Maxim series
Sky/Skyline series	Regal series
Super Majestic series	Roc series
Universal Plus series	
Ultima series	

August 1998.

Growing Pregerm Pansies

by Lisa Lacy

Pregerminated seed can give growers nearly 100% usable seedlings, essential for maximizing space and production. And thankfully, technology has brought us a seed that gets around the main drawback of early pregerminated seed—short shelf life. Now there's no need to use it within weeks of treatment.

Novartis Seeds, Downers Grove, Illinois, which developed PreMagic pregerminated seed, recently introduced PreNova seed. PreNova offers only a slightly lower level of germination than PreMagic (92 to 95% usable plugs), but it can be stored for up to six months, greatly increasing your sowing flexibility. It's as easy to handle as any primed pansy seed. Novartis Seeds holds two United States patents on this technology.

One of the popular PreNova products is Delta pansy, which is early flowering in the spring, heat and humidity tolerant for fall production, and quick to rebloom in winter low-light areas. If you're thinking of trying some of this new pregerminated product, here are some cultural tips that will help you get the most out of it.

In the Sowing Room
Handling
PreNova seed should be stored at 40° to 50°F. This type of seed will tolerate freezing, but avoid freezing temperatures if possible. Always sow the oldest seed first, and follow the expiration dates on the vials.

Delta pansy

Before sowing
Remove seed from refrigerator about two hours prior to sowing. Leave the caps on the bottles, and allow the seed to come to room temperature. Warming before sowing prevents the cold seed from sweating and attracting moisture from the air while in the seeder tray. Pansies are very sensitive to high soluble salts, so you may wish to test your medium before sowing or planting. A good way to tell if the medium has a high salt level is if the pansy radicle twists and grows along the soil surface and will not penetrate the medium.

To sow
Any type of seeder will work with PreNova seed: drum, needle, or plate. But never allow the seed to sit open in the seed hopper or in the greenhouse for long periods of time.

Moisture and covering

Because PreNova is pregerminated and then dried in the manufacturing process, the soil should be slightly moist before sowing. Here's a good rule of thumb for moisture management: On a scale of 1 to 100, where 1 is dry and 100 is total saturation, keep trays at 45. Avoid excess watering and excess covering with vermiculite or soil mix.

After sowing, cover with a thin layer of coarse vermiculite or 50-50 mix. Pass trays through the watering tunnel and into a chamber where relative humidity is around 85% and temperatures are at 62° to 68°F. Soil pH level should be 5.5 to 5.8.

Because PreNova is pregerminated, think of seed as starting at Stage ½. Water should be applied in a mist form during Stage 1. This provides the highest humidity around the seed without covering the seed with free water.

Summer Sowings

PreNova pansies should be held in a cooled area until radicle emergence. They benefit from two to three days in the chamber, but keep a close watch on the product to avoid stretching. PreNova seed trays may need to be removed from the chamber a day earlier than primed seed and two to three days earlier than raw or indexed seed.

Remove to benches, and place trays in a shaded greenhouse. If shading isn't possible, provide adequate misting to keep seedlings cool. When plants are big enough to handle, roughly three weeks after sowing, transplant to the final container.

Growing On

Temperatures

You should keep pansies at 60° to 62°F until about thirty to forty days after sowing, when the temperature may be dropped to 49° to 59°F. Night temperatures below 55°F may delay flowers. Higher night temperatures speed flowering.

Water

Avoid overwatering; a slight wilt between watering is OK. Feed with 100 to 200 ppm nitrogen in the form of calcium and potassium nitrate. Provide additional minor elements unless they're already incorporated into the medium at planting.

Light

Avoid direct sunlight after germination. Light quality and accumulation are the key factors in flower time. There's a dependency on light accumulation to initiate flower buds, and the quality of light is very important. Delta pansies have been bred to flower under lower light conditions and require less light accumulation, which is what gives the series its earliness.

Growth regulators

Growing cool and slowly will provide the best overall plant habit. Response to growth regulators varies per cultivar and color. Don't apply any growth regulator once buds have appeared, as this will reduce flower size.

There are a variety of chemical growth regulators that work well on pansies. Read and follow all label directions. The best nonchemical growth regulator on pansies is DIF, which will help control stem length and also hasten flowering.

Culture Notes, November 1998.

Growing Fantasy Petunia Plugs

by Dr. Dave Koranski

The new milliflora class of Fantasy petunias from Goldsmith has some excellent qualities that differentiate them from multifloras and grandifloras. Plants are very compact and free flowering. Flowers are smaller than multifloras (1½ inches in diameter) but are earlier and bloom continuously throughout summer with good heat tolerance. They make an excellent display in containers, baskets, porch pots, and hanging pouches.

With Fantasy, as with other new crops or varieties, you may need to change some environmental and cultural practices to get enough shoot and root growth in the plug stage. Because of their compact growth habit, root and shoot growth may be slower depending on environmental conditions. With a good root system, time to flowering is quicker than other petunias.

Generally, vegetative growth can be promoted by using more ammonium-nitrogen feed instead of nitrate nitrogen and more phosphorus. Photoperiod will also control flower induction and initiation and can sometimes interact with temperature to promote or delay flowering. Container size will influence shoot-to-root ratio. When this ratio is high (favoring shoot growth), plants tend to stay vegetative. Many times when plants are stressed with low moisture, more nitrate nitrogen, high light, temperature, and small container size, they will bloom faster.

Experiments to determine Fantasies' crop timing reveal they can be grown in any plug tray size, but to establish an optimum root system, you need a larger plant. The 288 tray produces a better plant than a 512 tray, because there is less stress on plugs. Roots should be active, and the root ball should be developed sufficiently before transplanting. Don't transplant without a good root system! This plant needs good root development to withstand stress.

In another experiment, plants fed with 20-10-20 at 100 ppm nitrogen for three to four weeks in the plug tray produced the best quality. Root and shoot growth were increased approximately 40 to 50% compared to using only 13-2-13-6-3 (N:P:K:Ca:Mg). Also, B-Nine was not necessary in the experiments.

Cultural guidelines

Water quality: Maintain low sodium and chlorides (less than 20 ppm).

Media: Starter plug media should be well drained and contain at least 20% perlite. pH should be 5.8 to 6.0. Soluble salts should be on the low side (0.5 to 0.7 mmhos, paste extract). Starter charge should have enough phosphorus (8 to 12 ppm). Maintain calcium levels of 80 to 100 ppm. During Stages 2 to 4, plugs can be fed with 1.0 to 1.2 EC.

Watering: During Stages 2 to 4, allow media to dry down before watering.

Nutrition: Fertilize with 20-10-20 at 100 to 150 ppm nitrogen during Stages 3 and 4. Supplement with 15-0-15 at 100 to 150 ppm nitrogen for one to two feedings to maintain calcium levels.

Temperature: Keep soil temperature during Stages 3 and 4 at 70°F day/62° to 65°F night. Warmer days than nights (positive DIF) will produce more vegetative growth. Avoid using negative DIF before mid-Stage 3, especially with small roots and shoots.

Lighting: Where supplemental HID lighting is used, early flowering may occur under long daylengths, resulting in a less than acceptable shoot-to-root ratio. Keep HID lighting to eight to ten hours per day until the plant is transplantable.

Culture Notes, March 1996.

Producing Cascading Petunias from Seed

The craze of trailing and spreading petunias from seed began with the popular Wave series and has continued with the introduction of the Trailblazer and Kahuna varieties this year. Retailers and consumers alike are clamoring for these seed-produced favorites, but, as with their vegetative cousins, producing these varieties requires a different touch than standard multiflora and grandiflora petunias. Following are some tips for success.

Trailblazer (left) and Kahuna (right) petunias

Seed

Most varieties come as pelleted seed to make handling and sowing easier. Sow in a well-drained, disease-free media in plug trays or open seed boxes. Although it isn't necessary, seed may benefit from a light covering of soil to maintain moisture levels.

Germination

Germinate at 72° to 78°F. Seed will require four to seven days to germinate.

Timing

Pink, Rose, and Lilac Wave will flower in twelve to sixteen weeks in spring and nine to thirteen weeks in summer. Purple Wave is four to ten days later.

A minimum of thirteen hours of daylength is required for shorter crop times. Trailblazer and Kahuna, according to the breeder, aren't daylength sensitive and will bloom in nine to ten weeks depending on the production location and time of year.

Temperature

Grow on in day temperatures of 65° to 75°F and night temperatures of 55° to 65°F.

Light

Provide 5,000 to 10,000 foot-candles of light. The more light, the better it is for plants, as long as you maintain moderate temperatures.

Fertilizer

These trailing seed petunias require slightly more fertilizer than typical hybrid petunias. Feed at every irrigation with about 200 ppm nitrogen. You can adjust fertilizer levels slightly up or down depending on light and moisture levels.

Growth Regulators

Because these varieties are vigorous, growth regulators can help control plant habit. B-Nine and Bonzi are effective. Frequent applications at lower rates will result in more uniform growth. For finishing plants, spray B-Nine at 2,500 to 5,000 ppm or drench with Bonzi at 2 to 4 ppm. For plug trays, cut rates in half. Stop applications at the first sign of visible bud.

Pests

Watch for whiteflies, thrips, and aphids. Pythium and botrytis blight can also attack these plants.

Varieties

Both Trailblazer and Kahuna, new for 1998, have well-branched habits. Kahuna is the more upright of the two, while Trailblazer is still slightly upright but spreads a bit more. Both have bright purple flowers that are just a shade lighter than Purple Wave.

The Wave series includes four colors: Rose, Misty Lilac, Pink, and Purple. While Pink, Lilac, and Rose are more upright spreaders; the original variety, Purple, spreads vigorously, making a carpet of flowers. Misty Lilac has the largest flowers of the series and grows slightly taller than Rose and Pink, with a spread that is between Pink and Rose.

Culture Notes, November 1998.

Salvia: Better Germ and More Plugs

by Dr. Dave Koranski

For plug growers to increase their profit margin, the one area that can make the most impact on the bottom line is to increase the total amount of usable seedlings that they can produce. This means getting the most germination possible from each seed lot.

For growers, moisture management is a key component of germination, because it relates not only to obtaining a high number of germinated seeds but also to achieving a high number of transplantable seedlings.

One of the difficult-to-germinate bedding plant crops that growers frequently complain about is salvia. Numerous articles have had some contradictory suggestions as to the amount of moisture that should be applied to salvia seed during Stage 1 and Stage 2. Therefore, we decided to conduct an experiment to determine the optimum moisture require-ments for salvia. We also looked at the effect of covering the seed and leaving it uncovered.

How Wet Is Wet?

First, it's important to describe the specific moisture terms used to describe the amount of water applied to a plug:

Wet/saturated: When the media is wet, water will ooze out from the bottom and top as the substrate is lightly touched.

Medium wet: The media is still glistening, but water won't ooze from the bottom and will penetrate only slightly from the top.

Medium dry: Media is not dark black and glistening, but it hasn't changed color to dark brown and virtually no water will ooze out of the substrate when touched.

Dry: The media has changed color to a light brown.

It's also important to recognize what a salvia seedling should look like in both Stage 1 and Stage 2 of plug production. The photos show both the plant tops and roots so that you can compare development of your own salvia seedlings.

Photo 1 clearly shows Stage 1 for salvia. The radicle is only slightly emerged, and these seedlings should be treated as a Stage 1 plant as defined

from the experiment. Photo 2, day eleven, shows seedlings that are clearly in Stage 2, where the moisture concentration of the growing substrate can be reduced, and seedlings can be fertilized.

Salvia - Salsa Red (Day 5)

Photo 1

Salvia - Salsa Red (Day 11)

Photo 2

Putting Salvia to the Test

We tested the effect of moisture and seed covering on the germination of salvia Salsa Scarlet by hand sowing seed into 288 trays first. For each treatment, four duplicate tests of one hundred seeds for each of four different seed lots were used. Lot 1 had a package germination of 100%; Lot 2 was 90%; Lot 3 was 96%; and Lot 4 was 94%. We tested seed lots with the plug medium both covered and uncovered for each of three moisture regimes: wet, dry, and normal.

Medium covering and initial watering-in was done with a top dressing machine and watering tunnel. Researchers maintained the moisture treatments for the first seven days after sowing. After that, production handled the trays under normal moisture levels as they would any plug trays in production.

Germination counts were made at fourteen and twenty-one days after sowing. An approximate usable plug count was conducted on day twenty-eight. The chart shows our results.

Treatment	Avg. 14-day germ	Avg. 21-day germ	Avg. usable
Uncovered/wet	91%	93%	85%
Covered/wet	82%	88%	77%
Uncovered/normal	98%	98%	96%
Covered/normal	96%	97%	93%
Uncovered/dry	95%	96%	90%
Covered/dry	98%	98%	95%

Don't Soak Them!

It's clear that the moisture level given during the first seven days of germination of salvia Salsa Scarlet has a strong impact on germination and seedling performance. Seeds germinated under a wet moisture regime show lower germination than those germinated under a dry or normal moisture regime. The usable plug counts, an indication of seedling vigor, are also reduced under wet conditions, especially for covered trays. For seeds under a normal moisture regime, performance is very similar for both covered and uncovered treatments. Under a dry regime, covered seeds have a somewhat higher usable plug count than uncovered.

What does this mean? Don't overwater your salvia plugs. Keep them at a normal or slightly dry moisture level, and if you tend to grow dry, top dressing your salvia plugs may help increase your germination percentages by a few points.

September 1998.

Seed Success with Fibrous Begonias

Recent research has shown that the highest germination rates for fibrous begonias are obtained by sprinkling seed evenly over media and germinating at 78° to 80°F, says Jim Nau in the *Ball Culture Guide.* If this is impractical, use  constant temperatures of 70° to 72°F throughout germination.

Plant one to two plants per four-inch pot, and five to six plants per ten-inch basket. Plants will germinate in fourteen to twenty-one days, and they take forty-five to fifty days from sowing to transplant. Grow on at 60°F.

Crop time is fourteen to fifteen weeks to sell green; for flowering packs sixteen weeks; in four-inch pots, eighteen weeks; and in ten-inch baskets, nineteen to twenty weeks. In the southern United States, allow fourteen to fifteen weeks for flowering packs and sixteen to seventeen weeks for flowering four-inch pots.

Pinching shouldn't be necessary for the dwarf varieties, but a soft pinch is good for taller varieties. They tend to grow upright, with minimal basal branching.

Plants will grow six to ten inches tall. Begonias will flower longer if planted in shade rather than direct sun. Also, green-leaved varieties will darken somewhat in full sun. In full sun locations, plants need to be established before weather turns hot, or they will be slow to develop and fill in.

Culture Notes, December 1998.

Plug Tips: Begonia and Verbena

by Dr. Roger C. Styer

At a GrowerExpo 2000 seminar in January, attendees had the chance to go one-on-one with plug expert Dr. Roger Styer for two hours to discuss their biggest plug challenges. Without a doubt, the plug crops most growers say they have a tough time with are begonia and verbena, so we asked Roger to offer production tips and ideas based on the successes and failures he's seen in greenhouses nationwide.

Begonia

Moisture management

Begonia likes it warm (75° to 77°F soil temperature) to germinate. It also likes to be kept uniformly moist until the first true leaf is about half expanded. The moisture level is very critical, due to the very shallow roots, until the first true leaf is out enough. At this point, the roots start to go deeper. That means you can't allow the media surface to dry off or turn a lighter color until that time. Failure to keep the media surface uniformly moist results in lower germination or uneven seedling growth.

Algae

If you do a good job with the moisture, you'll probably have problems with algae. Try not to compact the media surface during flat filling and seeding. And whether you irrigate by hand or boom, use a fine nozzle or nozzles to water the trays during the early stages. Keep the floors and benches clean of algae, and don't reuse plug trays for this crop.

Nutrition

Begonias don't like a high starter charge in the plug media, particularly NH_4. However, once you see some green leaves, you need to start feeding at low rates, such as 50 to 75 ppm nitrogen (N) from 20-10-20, 17-5-17, or 15-5-15. Once you can dry off the media surface, you can increase the feed to 100 to 150 ppm N and work in some 13-2-13 or 14-0-14. Northern growers will use 13-2-13 more often than 20-10-20. Alternate feed with clear water.

Getting begonias past the germination and early seedling stage with high germination and seedling uniformity is the hardest part. Pay attention to your moisture management, fertility, and algae control. HID lights will speed up fibrous begonia growth, whereas lights and long days are needed for tuberous begonias to prevent formation of tubers. You can grow tuberous begonias similarly to fibrous, except they take longer to grow and need long days.

Verbena

Moisture management

The main trick to getting good verbena germination is proper moisture management right from the start. This means giving verbena about *half the moisture* you give to any other crop. Covering the seed with coarse vermiculite will help get air to the seed while still holding humidity. Keep the moisture in the covering, not through the rest of the cell. Adjust your watering tunnels and watering-in techniques on the bench. Keep the trays on the dry side (but not bone dry!) until all the sticks are up, then you can go to a normal wet/dry cycle. You need to know where your verbena trays are during germination, and check on them three or four times a day. As far as temperature for germination, verbena like it from 72° to 78°F. Remember, do not overwater during germination!

Nutrition
Verbena is not a high-feed crop. Keep the media EC on the low side, and don't overfeed. Varieties have different growth rates, so exercise caution when using chemical growth regulators, as stunting could occur on slow-growing varieties.

March 2000.

Crop-by-Crop Checklists
by Dr. Roger C. Styer and Dr. Dave Koranski

Producing quality plugs requires that you know how key environmental and cultural factors affect particular crops. You must also check these factors regularly to avoid common problems and improve the chances of growing a high-quality crop. You should also know the four plug growth stages:
- Stage 1. Germination of the seed and emergence of the root
- Stage 2. Emergence and opening of the cotyledons
- Stage 3. Development of true leaves
- Stage 4. Hardening and toning the plug.

Here are our plug production checklists for eight of the most important bedding plant crops.

Begonia
- Growing media pH should be 5.5 to 5.8.
- Plants need light during the germination stage (10 to 100 foot-candles).
- The EC of growing media should be 0.7 for Stage 1 and 1.0 to 1.5 for Stage 2.
- Don't water in seedlings in Stage 2 with fertilizer with an EC greater than 1.5.
- The requirement therefore is low, constant feeding. Ammonium fertilizer is beneficial throughout.
- Growing media should be uniformly moist throughout Stages 1 and 2 and into early Stage 3 until the first true leaf is half expanded. At

that time, roots are going deeper into the cell, and you can reduce moisture levels to allow soil surface to dry out before watering again.

- For pelleted seed, pellets should be moist at all times until the radicle emerges, but the substrate shouldn't be saturated.
- The soil temperature requirement throughout Stages 1 through 3 is 72° to 75°F without lights or 70°F with lights. Warmer temperatures can cause poor uniformity.
- The calcium-to-magnesium ratio should be 2:1. Poor root development and growth structure can develop if calcium is deficient.

Begonia

Impatiens

- Growing media pH should be 6.2 to 6.5 from sowing the seed to selling the plants.
- Light is beneficial for days one and two of germination (10 to 100 foot-candles).
- Maintain uniformly moist conditions for Stage 1, and reduce moisture levels during Stage 2 to allow roots to penetrate the soil. Maintain a soil temperature of 73°F for both stages.
- Oversaturation of growing media in Stage 2 will reduce oxygen levels and damage root hairs.
- Growing media should have low EC levels of 0.4 to 0.8.

- Malformed leaves and stunted seedlings can form if EC, moisture, temperature, or pH is incorrect or if the seed has poor vigor.
- Shoot tip abortion can occur: if growing media EC is too high (1.2 to 1.5); if shoot tips are left wet for more than four hours (the growing point drowns due to lack of oxygen); if cold (lower than 65°F), saturated growing media causes a toxic buildup of ethylene within the plant; or if low pH causes sodium toxicity.
- Feeding calcium and reducing ammonium will produce a harder plant.
- Supplementary light for more than two weeks during Stages 2 and 3 can result in bleaching or yellowing of leaves (photo-oxidation).
- Leaf tip decaying quickly over a twelve-hour period indicates pseudomonas infection. Removing plants is essential.
- Reddish leaf spots indicate alternaria. Have a sample positively identified by a lab or clinic.

Marigold

- Growing media pH should be 6.2 to 6.5.
- If lower leaves are showing yellowing (chlorosis) or marginal browning or burning (necrosis), the cause is likely to be iron, manganese, or sodium toxicity caused by the pH dropping to 5.5 to 6.0.

- Minimize stress on French marigolds to prevent premature flowering by not allowing media to get too dry, by feeding with both ammonium and calcium fertilizers, by avoiding high light levels (greater than 2,500 foot-candles), and by not holding plugs past the transplanting date.
- Bud abortion and lack of flower development is generally caused by allowing the media to get too dry, by a lack of calcium, by too much ammonium fertilizer, or by excessive chemical growth regulators.
- Soil calcium levels should be around 120 to 175 ppm.
- Excessive leafy growth is usually caused by growing media that's too moist and ammonium levels that are too high.

Pansy
- Growing media pH should be 5.8 to 6.0.
- Non-uniform germination will occur if the growing media is too dry in Stage 1 or too wet in Stage 2.
- The EC of the growing media should be less than 0.5 for Stage 1 and less than 0.75 for Stage 2.
- An ammonium concentration of more than 5 ppm will cause seedlings to stretch.
- If the whole plant is getting leggy, the total nitrogen concentration, especially ammonium, in the growing media is too high, or the media is staying too moist too long.
- Average daily temperature for good growth is 68°F.
- Tip abortion or blindness can be caused by boron levels lower than 0.5 to 0.7 ppm in the soil or lower than 50 to 60 ppm in the tissue.
- Low calcium levels in the tissue (less than 1.75%) will produce malformed, puckered leaves. Calcium nitrate can be used as a feed, and calcium sulfate could be added to the growing media before sowing. The leaf symptoms may also represent the early stages of boron deficiency.
- Malformed, hard, leathery leaves develop with too high of a chemical growth regulator application (B-Nine at 5,000 ppm), especially at high temperatures (greater than 90°F).
- Thielaviopsis (black root rot) may develop as black lesions on the root when the pH is greater than 6.5. Preventative fungicidal treatment would be advisable. The disease spores can be spread in dust. Stressed plants grown under high temperatures are affected the most.

- You could try growing Tropicana vinca as an indicator plant in the same growing media alongside the pansies. At a pH of 7.0, it will quickly turn yellow to warn you of a problem.

Petunia, Grandiflora and Multiflora

- Growing media pH should be 5.8 to 6.2.
- Plants need light during the germination stage (10 to 100 foot-candles).
- Non-uniform germination and early seedling growth can be caused by saturated growing media in Stage 2. Adding 25 to 50 ppm potassium nitrate before sowing may enhance germination.
- Boron concentration should be 0.5 to 0.7 in the soil and 40 to 60 ppm in the tissue to prevent tip abortion and blindness.
- Strapped new leaves indicate a calcium deficiency. This can be improved by a foliar feed of 100 to 150 ppm calcium nitrate. This symptom could also be an early sign of boron deficiency.
- Elongation of the leaf petiole usually indicates excess ammonium or too-moist conditions.
- Yellow upper leaves are usually caused by high pH (greater than 6.6), which inhibits iron uptake.

Multiflora petunia

- To initiate flowering, plants need both soil temperature above 60°F and 13.5 hours of light per day. Without these conditions, flowering will be delayed.

Salvia
- Growing media pH should be 6.0.
- Plant stand is usually caused by too much wetness in Stage 2.
- Growing media EC levels greater than 1.5 in Stages 1, 2, and possibly 3 will cause stunted growth. Feed in the later stages of development.
- Ammonium concentrations greater than 5 ppm can be toxic. Symptoms show as a gray cast or browning on immature leaves in Stages 2 and 3.
- Interveinal chlorosis with veins starting to yellow on mature leaves indicates a lack of magnesium.
- Downward cupping of the leaves indicates that media is too dry, EC level is too high, or temperature is too cool.
- Salvia can be grown alongside other crops as an indicator of excessive salt concentrations in growing media.

Snapdragon
- Growing media pH should be 5.8 to 6.0. pH above 6.5 could cause iron deficiency, showing as interveinal chlorosis on immature leaves.
- Keep growing media EC levels lower than 0.5 to improve germination, control growth, and prevent some diseases.
- Tip abortion or blindness can be caused by: 1) high

Snapdragon

moisture on the meristem combined with cool temperature or high light; 2) fertilizer with EC > 1.0.

- Spotting on the upper leaves could be either downy or powdery mildew.
- Less than thirteen hours of daylength will promote vegetative growth and delay flowering for garden snapdragon varieties only.

Vinca

- Growing media pH should be 5.8 to 6.0.
- Darkness is needed during the germination stage.
- The soil temperature should be 78°F during Stage 1 and 72°F throughout the rest of the crop cycle until they are sold.
- Use a preventative fungicide against thielaviopsis (black root rot) and continue to inspect roots for black lesions.
- Damping-off at the surface of the soil is usually caused by rhizoctonia.
- Keep the media EC less than 1.0 throughout all stages, and allow the media to dry out between waterings for best control of growth and diseases.
- Upward cupping of the leaves would indicate the growing media is too dry and light levels are too high in Stages 2 and 3.
- Downward cupping of the leaves would indicate the temperature is too cool or the DIF is too negative.
- Roots not actively growing after transplanting could be caused by too low of a soil temperature; thielaviopsis; too high a salt concentration in the growing media (EC should be less than 1.5); or too high of an ammonium level (keep less than 10 ppm).
- Interveinal chlorosis on younger leaves would indicate the pH is too high (greater than 6.5).
- Tropicana vinca will quickly show chlorosis at high pH and can be used as an indicator plant for other crops.

From the book Plug & Transplant Production, A Grower's Guide, *published by Ball Publishing, copyright 1997.*

Winter 1996.

Chapter 6
How Do They Do It?
Plug Grower Profiles

⸎

One-on-One with Five of America's Best Plug Growers

by Chris Beytes

Plugs. Twenty years ago the word wasn't yet a part of our vocabulary. Visionary growers in Florida, Michigan, and New Jersey who recognized the plug's potential were just beginning to develop the tiny, singulated seedlings into a useful product.

Ten years ago the expanding availability of the revolutionary new plug forced bedding plant growers to begin analyzing the pros and cons of buying them in or growing their own. Millions, then hundreds of millions, were sold each year to a growing army of growers convinced to leave the seed sowing to the experts.

This year tens of thousands of bedding plant growers worldwide will buy billions of plugs from a few dozen plug specialists, making them one of the most important products floriculture has ever seen.

That rapid growth has yielded rapid change, both within the plug business and throughout floriculture. Greenhouse technology, communication systems, the needs of the ever-changing customer base—the plug specialist has encountered more change, in a shorter time, than any other segment of the industry.

Recent shakeups within the plug segment attest to this. Giant producers change broker alliances or ownership. Smaller plug specialists race to fill the gaps in production. Some, sharply focused on growing small quantities of superior quality plugs, suddenly find they've gained nationwide notoriety. Producer/broker alliances become like an exclusive club. If you aren't on the inside, you can't play the game. How are today's plug specialists coping with all the changes? And what new challenges are they preparing for?

The Changing Plug Customer

Plugs came into the industry at about the same time as Kmart and Wal-Mart. (It's a good thing—bedding plant production couldn't have kept up with mass-market demand without them.) Bedding plant growers divided into two camps: Those who produce for the mass market, and those who produce for independent retail sales. Both camps were eager to try the new plugs. Demand outpaced supply. They took whatever they could get, giving license to some plug producers to sell anything they could grow. Quality varied from week to week, variety lists were sparse, and service was practically nonexistent.

"Maybe five years ago you didn't have to listen [to the customer], because it was easier then to find enough customers who didn't specify [varieties] or didn't force you into meeting that demand," says Gerry Raker, who, along with his brother, Dave, and nephew, Tim, operates C. Raker and Sons, Litchfield, Michigan, one of America's most respected plug suppliers. "I don't think it's that customers didn't want that attention. I just don't think they got it."

Ron Wagner of Wagner's Greenhouses, a Minneapolis, Minnesota, plug specialist, agrees. Along with his parents, Rich and Nola, and brother, Scott, Ron has adapted to tremendous changes in the industry over the last several decades. He says customer service, especially on cultural issues, is more critical today because of cutbacks in university extension services and because many brokers don't have the technical service departments they've had in the past. "I think more [plug growers] are realizing that it's not just being able to grow a product; it's being able to support it," he says.

Gerry Raker, C. Raker and Sons

Gerry calls it value. "A seed is a seed, and media is media, and a plug is a plug," he says, "so where does the value come from? It's matching [your product] to what the customer wants to do with it and what he's going to make with it."

Today's plug customers are getting plenty of attention, support, and value thanks to Gerry, Ron, and others who early on recognized the importance of meeting their needs.

Selling Service

Exactly what service entails depends on the customer. For high-volume plug buyers, price, not selection, is the driving force. The increasing number of large growers using automatic transplanters also drives a demand for more uniform plugs and higher counts of usable plugs.

Dave Tagawa of Tagawa Greenhouses, Denver, Colorado, one of the nation's largest plug producers (thirty acres at four Denver sites), adds that large growers are also demanding more communication between the producer, the broker, and themselves. But large growers have been least likely to buy plugs, especially as their own competition heats up. Most feel they can cut costs or increase their control over the finished bedding plant by growing their own plugs. That leaves most plug specialists focusing on small- to mid-sized growers.

David Wadsworth, owner of Suncoast Greenhouses, Seffner, Florida, started his plug business in the 1980s in the same way as many other plug growers: as a finished bedding plant grower producing his own plugs. He says feedback from his contacts in the business about both the high quality of his plugs and the dissatisfaction with large plug growers, who seemed to have lost sight of the needs of small customers, convinced him he had a future as a plug specialist. And as a small grower himself, David knew the importance of delivering what he promised and, just as important, letting his customer know when he couldn't.

"I came in [the plug business] understanding that every single tray on our bench counted, that it belonged to somebody," David says. "I had to be right on top of accounting for every single tray that was out there on a bench."

Accounting for every tray and communicating that tray's status to both the broker and the end user are two of the biggest challenges facing plug growers. After all, adds John Miller, Tagawa's manager of prefinished material, "If we're not able to give a customer 100% of what he ordered eight months ago, their question is, 'Why?'" Plugs are a highly seasonal business, with peak production taking place in just a few months of the year. Maintaining the large variety list and wide range of plug sizes that today's customer expects greatly complicates the job of tracking inventory.

David Wadsworth, Suncoast Greenhouses

Plus, most growers are moving their plugs almost constantly, from the sowing room to germination chambers and through Stages 1 to 4. During the short six weeks or so that they have the plug, they've got to check for germination percentages and usable plugs, patching trays at some point in the process. "It's a daunting task if done properly, and one that's easily under-estimated," says David.

Most of the improvements plug specialists are making involve invest-ments in logistics and communication. Gerry Raker has been a pioneer of using bar codes to track plug tray inventory through the system from sowing to shipping. Green Circle Growers, Oberlin, Ohio, has also been using bar codes to precisely track their inventory. They've been developing a bar-code-driven sorting and shipping system to help with another challenge: getting their product out of the greenhouse and to the customer. Ron Wagner says he'll be upgrading to a bar-code-based system like Raker's within the next year. Right now he uses a computer-based system that's effective, but the upgrades will make it even more efficient.

However, some growers are looking for the next step beyond basic bar codes. David Wadsworth sees logistics—both in the greenhouse and in distribution—as "the challenge that's going to separate plug suppliers now and even more so in the future." He's currently investing much of his time looking outside of our industry to find better ways to manage his wide variety list and provide maximum service to his customers.

"I don't know how carefully the larger operations and the newer expansions that are going up now are addressing these issues," he says. "I have yet to visit a large-scale operation that, in my opinion, really has a complete handle on their inventory. They do large volume, but as to how well they can account for every single tray out there that belongs to someone . . . I think there are very few who can really claim to be on top of that."

David's internal logistics solution may come from the warehousing industry. "We're not that different in our logistical problems," he explains. "There are some really fascinating solutions that have been developed in that field. We're managing inventory on two dimensions: x and y. The modern warehouse is managing inventory on x, y, and z—on three dimensions."

Regional or National?

Thanks to efficient air-freight systems, nearly all plug specialists ship their product coast to coast. But more and more plug growers, searching for a competitive advantage, are now offering special services such as rack delivery (which has to be done via truck) and same-day or overnight service right to the customer's greenhouse. Wagner's customers who have ordered plugs can get them through virtually any shipping system, but thanks to Wagner's central Minneapolis location, some customers pick up their plugs at Wagner's dock, a benefit other plug specialists in less accessible areas can't offer.

David Wadsworth is another strong proponent of such quick, local service. "Once customers have seen the benefits and had the experience of product delivered to their door, soon after that they don't want to talk to you about having to go to the airport to get it or having it shipped to them in boxes that have to be unpacked and later disposed of."

Still, he and others admit that there's still a push to cover the largest market area possible to gain the maximum customer base for their product. Plugs are, and always will be, a high-volume business.

To move these high volumes of plugs, Tagawa Greenhouses is adapting the cutting edge in delivery systems. This is important because they're behind other plug competitors when it comes to greenhouse efficiency and labor-saving automation. While most plug specialists now grow in dedicated plug ranges featuring Dutch tray systems and boom irrigation, much of Tagawa's facility is older cut flower greenhouses with rigid benches. They use extensive hand labor throughout.

Left to right: George Tagawa, John Miller, Jim Tagawa, son-in-law Chuck Hoover, and Ken Tagawa, Tagawa Greenhouses

One way for them to gain an edge is to cut shipping costs to customers. They've done it with a new pallet shipping system that Randy Tagawa, company general manager, first saw being used in Europe. Using custom Blackmore shuttle trays and wooden pallets, Tagawa can ship 160 plug trays (equivalent to 40 cases) one way to customers. This eliminates shipping on metal racks that have to be returned, adding to shipping costs. After three or four loads, growers send back the shuttle trays and pallets for recycling. If 50% of the material is recycled, Dave estimates the new system will cut their shipping costs by 40%, giving them a serious competitive advantage in one area.

Broker or Direct?

The relationship between a plug specialist and his broker(s) may be the most important—and the most tenuous—in floriculture. Ken Tagawa, CEO of Tagawa Greenhouses, likens their exclusive ten-year-long relationship with Ball Seed Co. to a marriage: "If you decide to get married, you're going to make it work. And you have to pick your partner right."

Bruce Knox also knows the importance of a good grower/broker relationship. With his Orlando, Florida-based Knox Nursery, he is a ten-year veteran of the plug business but is coming into it this year on a major scale, having recently completed construction of a six-acre, $6-million facility. When first getting into the plug business, Bruce says he "seemed to be on the outside looking in" at the relationships between big plug growers and major national brokers, with no way to get a foot in the door. Now he has a solid and expanding relationship with several major and regional brokers and plans to work hard to maintain those relationships. As Dave Tagawa puts it, "You can't just walk in [to a broker's office] and ask, 'Do you want to sell my plugs?'"

Is it better to have just one broker partner, like Tagawa, or several, like Knox, Raker, Wagner, and Suncoast? What about going direct and bypassing the broker altogether? Arguments support the first two, and despite fears of brokers, no plug grower we talked with sees any future in the third option. They all agree that brokers provide an invaluable sales function that they wouldn't want to have to do themselves.

Dave Tagawa likes the "focused direction" that comes from working with one broker. "I see difficulty in dealing with multiple broker relationships because of the different systems that have to be maintained and the different types of information that you have to deal with," he says.

Bruce Knox, and father Jim, Knox Nursery

Ron Wagner works with multiple brokers but admits, "It's harder to play on more than one team." Gerry Raker also works with multiple brokers but says the broker has to add value to his product. "We don't want to do without brokers," he says, echoing the feelings of the others. "I just don't want a lot of brokers listing our stuff. I want someone who's going to sell my product—present it, sell it, follow up on it, present all the features and benefits."

Embracing Technology

EDI (Electronic Data Interchange), while hardly a new technology, can take floriculture into the next century—if we can somehow come together to develop the standards required to make it work.

EDI allows real-time communication between plug growers, brokers, and customers concerning inventory, order status, availability, etc. Anything that's in a company's computer system will instantly be available to business partners. The mass market already requires EDI links with most of its vendors. A few suppliers and brokers like Wagner's and Vaughan's Seed Co. are already linked via EDI, although it's still in the testing stages and isn't regularly used yet.

Overall, our industry has been slow to adopt this new technology, frustrating plug growers who see it as vital to their businesses. A group of brokers and suppliers has been meeting to develop EDI standards for our industry, but it's been a slow, frustrating process, in part because it requires each business to put aside its own individual way of doing things in favor of the community good—not something for which our industry is noted.

For their part, plug growers and other suppliers are more willing than brokers to move on this new standard, Gerry says. And each has already been making major investments in computer hardware and software technology. Ron Wagner now has an MIS (Management Information Systems) employee on staff to coordinate his extensive computer network, which includes fourteen PCs running Windows '95, a Unix server, a new Windows NT server, Internet access, and EDI. Raker's has helped get PCs into the homes of 80% of their full-time employees to help them become more comfortable and experienced with computer technology.

Man Versus Machine

When you're dealing with a product numbering in the tens or even hundreds of millions, efficiency is critical to profitability. The early years of the plug industry were spent focusing on culture; today's big push is toward automating to cut costs and increase production.

Bruce Knox owns America's newest plug range. His first plug trays were sown in early February. The six-acre facility features all of the bells and whistles needed for growing a good plug (see "Maximum Design" page 170). Are facilities like this upping the ante for other plug specialists? There's no easy answer, but all agree that it's made it harder to come into the business.

Ron Wagner recently got a $500,000-plus quote to finish completely automating his seeding lines and table handling equipment in the sowing room. "Is that cost effective for that six-week peak in labor for our size facility [4½ acres]?" he asks. Obviously, expensive equipment makes more sense for large operations, but for Ron, investing in a bar code system and continuing with more cost-effective improvements to his seeding lines are necessary and profitable choices.

Ron Wagner, Wagner's Greenhouses

Gerry Raker takes a more philosophical view. "Automation to me is a negative term," he says. "Building efficiency through people is positive. Combining people and technology is positive, but trying to build automation to replace people doesn't make sense."

Raker's major focus recently has been on completely redesigning the organization of the company, disregarding concepts such as departments and employee job descriptions. The end result will be a company more able to quickly and efficiently respond to change. (See "Zeroing in on Plug Production Details" on page 178.)

Change or Face Extinction

Industry experts estimate that plug specialists sell just 30% of the plugs used in North America. The other 70% are produced by growers for their own use.

Plug growers see a lot of plug business still to be had. That's one of the things that keeps plug growers going despite the continuous changes they face.

"Our assignment," says Gerry Raker, "given the fact that quality and service are a given, is to be able to change and learn as quickly as things are happening. That's the key to profitability. If you get locked into a production method or style or program and the market changes . . . well, that's how the dinosaurs became extinct, right?"

April 1997.

Maximum Design

by Chris Beytes

Bruce Knox knew it was time to build a new plug range. After carefully evaluating his company's direction and goals, the Orlando, Florida, plug and bedding plant grower decided to put more emphasis on plugs, and that meant updating from their old, inefficient structures to a facility that would put them on a level playing field with other established plug growers. The end result is a spectacular new six-acre range that incorporates the best of everything with an emphasis on quality of growing environment and efficiency of movement, two factors critical for competing in the ever-changing plug market.

Knox Nursery has grown plugs since 1984, when they started growing them for their own finished bedding plant business. A cash crunch after building a 50,000-square foot sawtooth, single-poly plug range pushed them into selling some plugs and forging relationships with a few independent brokers and then eventually with national broker Vaughan's Seed Co. Plug volume went from 7 million units in 1987 to 50 million units and 50% of their total business by 1995, when Ball Seed Co. started selling their product. In 1996, plug sales jumped to 70 million units and 70% of their total business. Bruce, president of the second-generation family business (brother Monty is responsible for finished production) anticipates that plugs will make up 80% of their 1997 business.

"We had to make a decision this year. Are we going to continue to grow, or are we going to scale back?" Bruce says of their plunge into plugs. "We can't do a good job pushing the volume through our old place. This past

spring, we literally had people tripping over each other." With a commit-
ment from Ball Seed to sell their increased production, Knox decided to
build a state-of-the-art, 261,000-square foot greenhouse about fifteen miles
from their old site.

Bruce first contacted Hove International (United States representative for
Alcoa/Intransit) about designing a flow plan for a plug range, incorporating
Dutch tables for maximum movement and space efficiency. Bruce took the
plan to Nexus Greenhouse Systems and asked Jeff Warschauer, vice president
of sales, to design a greenhouse around it.

One very unique feature of the job was the way Jeff brought all the
vendors together early in the planning process to offer their input about the
project. Representatives from L.S. Americas, Wadsworth Control Systems,
Argus Controls, Biotherm, and Hove International met several times with
Bruce and Jeff both on and off site to discuss how to integrate their own
systems with all of the others to best meet Knox's needs.

The Nexus retractable roof greenhouse (shown nearing completion) is built
with the same design as the rigid house, allowing easy conversion to a
permanent roof.

The end result is a showplace of modern greenhouse design and efficiency.
The backbone of the range is a Nexus greenhouse featuring 42 foot 8 inch-
wide bays that are 300 feet long. Bay width is designed to accommodate three

rows of Intransit tables. Gutters are thirteen feet six inches high, the maximum Orange County would allow without extra engineering for wind resistance. Covering is clear Dynaglas for maximum light transmission and energy efficiency. A Wadsworth curtain system equipped with Ludvig Svensson shade fabric (50% shade) doubles as a heat-retention system.

Fan and pad cooling is a must in Florida just to keep greenhouse temperatures close to outside temperatures. Interestingly, Jeff mounted the American Coolair pad system high along both sidewalls of the house, and fans are mounted in the roof over the center aisle. Bruce says changing the environment up high instead of at bench level keeps the greenhouse temperature more uniform. HAF fans keep temperate air circulating.

Hot water heating (yes, you need heat in central Florida; Bruce runs it about four months out of the year) is provided by nine Hamilton boilers feeding underbench and overhead pipes. The entire range is monitored and controlled by an Argus Controls computer system.

In the 30,000-square foot headhouse, the sowing room is fully automated with two lines of Flier equipment: tray destacker, label printer, filler, dibbler, vermiculite topper, and water tunnels. Sowing is done with SK Designs drum seeders or Hamilton needle seeders, depending on seed type. Bruce says the two lines should easily handle 500 trays per hour, his expected output.

The greenhouse is laid out in three zones corresponding to plug growth stages. Stages 1 through 3 account for four of the six acres. Trays flow from the sowing area into the 85-by-100-foot germination rooms (equipped with automated Intransit table stackers) or directly into the greenhouse. Like most Dutch table ranges, trays travel from zone to zone in aisles at the perimeter of the range. An extra aisle along the center of the 300-foot wide bays increases flexibility, even if it does cut down on usable space. Plug trays will be tracked with bar codes; a local company is developing the software.

The most unique aspect of the entire range is Stage 4: a two-acre Nexus retractable roof greenhouse, Nexus' first. Serving as a rollout area, tables slide out through a wide opening in the side of the rigid roof greenhouse into the retractable roof house where plugs are hardened off. The retractable roof house is identical in design to the rigid house; just move a few perlins, and it can be skinned with rigid covering. It's equipped with both a waterproof covering and 50% shade fabric, allowing maximum environmental flexibility year-round.

Irrigation in Stages 1 through 3 is handled by ITS booms and hand watering. Each boom covers three rows of tables, but three solenoids let Bruce water tables in any combination. Chemicals can also be applied through the booms. The retractable roof area isn't equipped with booms—yet. Bruce admits they ran short of money, adding that they still aren't sure of the best way to irrigate this section. Another job they're still not sure how to best accomplish is shipping. They'll go through a season at the new range before making any investments in this critical area.

Going from their old 110,000-square foot range into the new facility will let Bruce increase his production to 200 million to 300 million plugs a year. That volume is important considering his investment: $23 per square foot total, including land—more than $6 million. Bruce admits it's a big investment considering how tough the plug business is but says the improved environmental control will let Knox grow plugs year-round, not just seasonally. Plus, he'll look at prefinishing or propagation as possible ways to use the facility during slow times.

April 1997.

Bar Codes Solve the Plug Inventory Dilemma
by Chris Millar

Knox Nursery's new state-of-the-art, six-acre plug range in Winter Garden, Florida, is definitely a showcase of the latest greenhouse technologies. In planning and developing this new greenhouse, Bruce Knox's goal was to take advantage of any technology that was available and practical that would reduce labor costs. From the seed sowing lines, which are capable of producing 1,800 plug trays an hour, to the high-tech Argus environmental control system, this theme is evident throughout the greenhouse. Automation and technology are used at every level.

One key area, however, lagged behind the rest: inventory control. While every other system was in place when production began more than a year ago, Bruce hadn't settled on a system for tracking inventory, which resulted in some fairly serious growing pains the first season.

That problem is behind us now, thanks to a bar-code-based system that tracks every table and every tray through every step of production, from sowing to shipping. And best of all, 80% of the system uses off-the-shelf software, saving us time and money. Our total budget was $40,000. Here's how it works.

Too Many Plugs, Too Much Movement

Knox Nursery produces almost 1,500 varieties of plugs in five different tray sizes. Most of the orders we ship out include several different plant varieties. One of the prominent features of the new plug range is the Dutch table system we grow on. Each table holds forty-two plug trays and can move easily throughout the greenhouse on a system of tracks. While these tables make it easy to move product they also present a liability: In a greenhouse equipped with Dutch tables, plants tend to be moved around considerably more than in a range that has fixed benches. Since it's practically an effortless task to move several tables, our growers take advantage of this to shuttle the tables into different growing zones in the greenhouse during each crop's growing cycle. However, this creates a challenge when it comes to keeping track of where the product is at any given time. The logistics of pulling multiple orders that all have dozens of varieties are complicated enough— pulling orders is even more difficult if you don't know where the plants are in the range! To put our situation in perspective, we currently have a capacity for 3,500 tables and 147,000 plug trays—and our future plans for expansion could more than triple our capacity.

The main challenge was to come up with a system that would track each individual plug tray from sowing all the way through production to shipping. We wanted to keep an accurate inventory of what quantities were on hand as well as the location of each plug tray.

Our first two seasons in the new range demonstrated a need to have a better handle on our inventory. We had used manual counting and verifying, which gave us fairly accurate numbers on what we had in the greenhouse. But that didn't tell us the specific locations of anything. We tried to group similar crops with the same finish times in the greenhouse, although at times these groups could consist of several hundred tables. To search through that many tables for one specific tray can be time consuming and tedious, to say the least. One of the most frustrating things that can happen to a plug producer is to spend several weeks growing a tray for a specific customer and then not find the needed tray at the time of shipping—only to have the tray in question show up weeks later, ready for the dump! We wanted to make sure that never happens at Knox.

Bar codes and scanners, both stationary and handheld, makes keeping track of the approximately 147,000 plug trays and 3,500 tables Knox has the capacity for easy.
Photo: Chris Millar

Our Requirements

We decided early on that bar codes would be at the heart of our inventory system. Bar codes are widely used today, and there is a variety of equipment that can read or scan them. We were already using custom-made label printer/applicators to label our plug trays; adding a bar code to our existing label presented no extra cost and was fairly easy to do. Plus, we found no practical alternatives to bar codes for identifying our plug trays, Dutch tables, and various growing locations in the greenhouse.

The first and most important step was to record what was being sown at the beginning of production. In seed sowing we didn't want to slow down the production lines in any way. If possible, we wanted to record what we produced in a totally "hands-free" fashion.

After each tray is sown, it's laid onto a table with an automated table loader. We first needed to associate each plug tray with the table on which it's placed. Then, as each table is moved into the greenhouse, we needed to be able to record its location. Plus, all table movement from this point until the product is shipped needed to be recorded as well. By keeping track of table movement, we'd know exactly where each plug tray was when it came time to ship it.

Prior to shipping, we wanted to be able to print picking tickets that include the products' specific locations. When a tray is ready to be boxed, we wanted to remove it from inventory. In addition, Bruce thought we should use the system as a final quality-control check, making sure each tray is actually needed for the order being boxed. Finally, we wanted to print our FedEx labels with the contents of the box printed on the label.

Warehousing Provides the Answers

Once our requirements were outlined, we began our search for ways to make it happen. The first place we looked was commercial inventory software designed for the horticultural industry. This software uses handheld wand- or pen-type scanners. This wasn't the solution for us, as handheld pen scanners would require too much labor. Next, we looked to warehouse inventory and "plant floor data collection" specialists for help. After all, in theory, our needs aren't that much different than a typical warehouse. Just like most warehouses, we store our product in containers that can be moved from one location to another.

We invited four companies that specialize in inventory and data collection to review our requirements and put together proposals on how to meet our needs. We received four very different solutions. Here's what we chose.

The Solution in Production

This application, especially in production, could be called "data collection." IBM makes a line of data collection terminals that are capable of collecting data from a bar code scanner. The backbone of this system is the IBM 7526 Data Collection Terminal. We have four of these connected to a PC using an Ethernet network. There are two of these small units in seed sowing—one on each of our two lines in production and another pair in shipping. Each of the terminals in seed sowing has two bar-code scanners connected to them. Both of these scanners are "raster" laser scanners, which are the best type for reading bar code labels that are moving. One scanner reads the labels on the plug trays as they move down the conveyor line. The other scanner reads the bar codes on our Dutch tables as they move out of the table loader machine.

The Solution in the Greenhouse

Every growing location (a row of twenty-four tables) in the greenhouse is also labeled with a bar code. As the tables leave seed sowing, we have a record of which trays are on which tables. To track the movement of the tables in the greenhouse we use Symbol handheld laser scanners. By scanning the table and its location, we now have the complete inventory of what was sown and where it is.

Managing the inventory in the greenhouse requires some discipline. We have to make certain that anytime there is table movement it's recorded with one of the three handheld scanners. Anytime there's any inventory shrinkage, it needs to be recorded with the handheld units. These handheld scanners record the information you scan until put in a docking station that uploads the information to the PC. We could have invested in Radio Frequency scanners that would automatically update the computer, but they're more expensive (although coming down in cost) and only save a small amount of time.

The Solution in Shipping

The first thing that's done at the beginning of a ship day is to load into the PC a list of all orders to be shipped. Next, we print picking tickets that detail each customer's order and the product's location in the greenhouse. Employees assemble these trays and push them up to the shipping warehouse for packaging.

We have two data collection terminals in shipping that each have two handheld scanners attached. There's a scanner located at each of the four packing stations that are used to scan the plug trays as they are being packaged. Once a tray is scanned in shipping, the first thing that happens is a check for accuracy. Does this plug tray belong to this order? Next, the tray is removed from the Inventory Database on the PC. Finally, we print the required shipping label, which also shows the box's contents.

As mentioned before, we have a network of four data collection terminals connected to a PC. The actual inventory is maintained on this PC using Microsoft Access, a popular database program. The total software solution was about 80% off the shelf and 20% custom written. IBM provides a program called DCC Connect with their data collection terminals. This keeps the network running and records what the terminals report. The custom-written software moves information from DCC Connect to Access, verifies the orders, and creates the shipping labels.

Benefits

The first obvious benefit of our new system is accurate information. We now know the exact quantities we have on hand, as well as the location of every plug tray, including the table it's on and the specific greenhouse location it's in. Plus, using bar codes to manage the inventory beats manual counting any day. We expect to see a big payoff this spring when we start shipping large numbers of trays, as the labor spent on pulling and assembling orders should be drastically reduced.

Customers benefit, too. By having a better handle on our inventory we can avoid last-minute cancellations. And by printing the contents of the box on the shipping label, our customers can identify the material they want to unpack first without opening every box in their order.

Advice to Other Growers
We experienced long delays with our programmers for various reasons. When entering into any type of contract specify a bonus or penalty for making or missing deadlines. Break your project up into sections that make sense. This may be easier than trying to tackle a big job all at once.

What's Next?
The next step for this inventory system is to tie the database into our company's mainframe. We have future plans to upgrade our main computer, as well as the software we use to run the company. When we're ready to do that, we'll certainly want to link the two systems together.

Knox even has a Web page under construction (www.knoxnursery.com) that will be linked to our inventory system. With a better handle on inventory we'll be able to post availability lists much earlier than at present, giving us more time to sell anything that's not sown for a precommited order.

And one other use for the system may be with finished bedding production. There's no reason we can't track flats and pots using the same bar code system, although with more hand scanning, bringing some of the same efficiencies to our finished plant business that we expect to realize in our plug business.

September 1998.

Zeroing in on Plug Production Details
by Don Grey

Successful plug production requires careful attention to detail through many stages, each building on the other to produce a healthy, salable product.

None is more critical than what growers at C. Raker & Sons, Litchfield, Michigan, call Stage Zero (0).

Stage 0 is the precursor to Stages 1 through 4. It includes everything that takes place in the seeding room—tray filling, seeding, watering. It also includes purchasing material, even keeping records of production lots for quality-assurance tracking. Each represents a critical Stage 0 step; a misstep here can spell problems later. "Stage 0 lays the floor for our entire operation," says Jenny Kuhn, Raker's Start Team leader. "It's very important."

The Start Team

Raker has been producing plugs for more than twenty years. The company has five acres of greenhouses and produces at least 100 million plugs annually. Although bedding plants represent its "bread and butter" business, production is moving more and more into perennials. About 30% of Raker's business is now in producing some 250 perennial varieties.

Over the past couple of years, Raker has redesigned its organizational structure from a traditional system to a team concept. It's taken at least a year or so to work out the new system, but it has greatly improved communications and workflow throughout all company levels, Jenny says.

Raker's Start Team is responsible for Stage 0 production. This is one of four teams comprising the Production Team, which itself is one of several company-wide teams (Physical Systems, Customer Service, and Technical Support are others). The other parts of the Production Team are Mid-Team, End Team, and the Grow-Flow Team.

The Start Team has four full-time, year-round workers, expanding to as many as twelve during the December–April and July–September busy months. Jenny says all of the seasonal workers the team hires receive in-depth training and many return season after season. Teams also interchange members as needed during peak cycles.

Team members are cross-trained so that no one person works only one task. Members switch tasks as often as hour by hour. This helps keep boredom down and interest and responsibility high. The team is not slowed if a member is absent because of illness or time off; others merely pick up the slack. "The majority of the team can work in any capacity," Jenny says. "Everyone is cross-trained. Everyone is knowledgeable of the specifications."

The Start Team holds weekly meetings. The entire Production Team also meets together, and meetings with the greenhouse's full work force ensure information flows back and forth. "People have the information necessary to make decisions," Jenny says. "It's part of our progression."

All about Details

Running a successful Stage 0 seeding operation requires a fine attention to detail. Work takes place in a multipurpose building that also doubles for tray and media storage and flat filling. Although Jenny would prefer a climate-controlled room, everything possible is done to keep the seeding operation as clean and free from dust, heat, and humidity as possible. Seeding is kept out of the greenhouse where humidity and sunlight can wreak havoc on sensitive seeding machines, and where seed can degrade if exposed for too long.

Purchasing

First, Raker's Start Team is responsible for buying many of the production components: seed, trays, media, and equipment. Purchasing is nearly a full-time job. Jenny spends most of her time purchasing seed, allowing the others to handle the production details. To get the seed she needs, she stays on top of the market. Staying in contact with seed company representatives is important; it enables her to spot shortages or good buys and order accordingly. Jenny will order seed in advance wherever and whenever possible, particularly for hot sellers, though a lot of what she buys is on an as-needed basis as contract orders shape up. Experience and "grower intuition" help her deal with the difficult varieties that produce late seed or have lower germ rates. This attention pays off, however, in the seed quality and supply that is critical for a plug grower.

Media

If seed is important, media can't be far behind. Good seed and quality media are the one-two punch of the plug business. Try germinating good seed in poor media and see what you end up with.

Of course, just any media won't do. Raker's uses two basic mixes: one for cyclamen transplanting and one as its basic plug mix (both of which are supplied by Sun Gro Horticulture). Indeed, the greenhouse works closely with Sun Gro to develop and maintain the media to their specifications and has a written supply contract that's tailored to their needs. "Know your media company!" Jenny advises. Because of its tight specifications, Raker's mixes come ready to use right out of the bulk bag, Jenny says, but calcium clay (7%) is added.

What does Raker's look for in media? Consistency and uniformity. Peat fiber length and uniformity are critical in producing a uniform plug tray. To ensure that quality remains consistent batch after batch, Raker's takes a media sample from each batch, and it's bagged, labeled, and stored for the duration of the production cycle. That way, if problems arise later, Raker's staff can go right to the sample to see if the problem is with the media.

Tray filling

Uniform media makes it easier to produce consistently filled plug trays. Raker uses four primary tray sizes: the standard 512 and 288 for most production of bedding annuals, and 338 and 128 for most perennial production. Raker occasionally uses other sizes, such as the 55 for some transplants and the 128 deep for some crops, like dracaena. It uses other sizes on a speculative basis or if a contract calls for them. Most of the trays are Blackmore products.

Trays are filled by a Bouldin & Lawson flat filler. They're filled only when needed and are no more than a day old. Freshly filled trays mean the media's moisture content stays fairly constant, and trays don't dry out too much. Moist media helps to hold the dibble, which in turn helps in proper seed placement. Proper seed placement is critical in producing a plug that grows full and even in the cell.

Before trays go into the flat filler, they're checked for holes in the bottom and for even edges. They're also inspected to make sure they aren't bowed and will sit flat. Even a slightly bowed tray can cause growing, uneven drying, and water problems.

How trays are filled is also critical. "It's not as easy to do as you think," Jenny says. "A huge component is to make sure they're filled uniformly and compacted properly. You don't want a half plug. Tray filling may seem to be such a minor thing, but the detail is critical."

As the tray cells are filled, they're rolled and compacted. A brush whisks away excess media, which is captured and reused. To maintain uniformity of filled trays, they watch the brushes on their tray filler for signs of wear.

Raker usually wants media compaction to be firm, with some give to it. Even compaction provides for uniform root growth. Almost by second nature, Start Team members can tell if a tray is packed too tight or too loose. As trays are filled, they gently check them with their fingers. Improperly filled, compacted, or dibbled trays are rejected.

Once filled and compacted, cells are dibbled to make a depression for the seed. It's important to center the dibble and the seed so the plug doesn't grow to one side or the other of the cell. "Proper tray filling affects your finished plug," Jenny says. "There's no doubt about it. It's got to be uniform."

Seeding

The previous careful attention pays off during seeding. Raker's Start Team uses three people on a seeding machine at all times: an operator, a feeder, and a catcher. This three-person crew checks and completes all the necessary paperwork, inspects trays for uniformity, checks the seed as it goes through the seeder, and monitors seeding to make sure each cell in each tray is properly seeded. "Bringing one more person in [to the seeding line] really helped," Jenny says. "The machine is only going to be as good as your operators." All three people rotate seeding machine tasks on occasion, sometimes as frequently as each hour.

Raker uses Blackmore seeders, topcoaters, and watering tunnels. The relationship with Blackmore is as old as the company itself, Jenny says. Raker's has two high-speed cylinder seeders, one needle seeder, and one turbo seeder. The turbo and needle seeders are used infrequently these days, usually for unpelleted tuberous begonia seeds. Jenny relies heavily on the cylinder seeders, even for difficult-to-sow seed, such as marigolds.

Every two hours or so, Start Team members will stop seeding to run test trays, checking to see that the trays and seeder are properly calibrated. A few trays are covered with wet paper towels and run through the seeder, which is filled with dead pelleted seed. It takes a little extra time, but the results show in fully seeded and uniform plug trays.

At the end of each production run, team members blow off machines by using an air hose. This helps prevent the next tray with the next crop from picking up unwanted seed.

For those crops requiring a topcoat, trays are fed through an automatic topcoater. A topcoat is usually a light dusting of media or vermiculite. "A uniform cover is very important," Jenny says. "You don't want too much soil."

It's important to handle the trays as infrequently as possible, Jenny says. Care must be taken not to jar the seed off-center through vibration. Indeed, Raker's is looking to automate even more of its tray production to facilitate tray movement with direct placement on benches. Less handling not only saves labor but also ensures the integrity of seed placement in the tray cells.

Watering

Seeded trays are run through a dedicated watering tunnel, which is adjustable for water volume and tray speed. Raker's rule of thumb is to water just to the point of a drip and no more. The one thing they don't want is nutrient leaching. Most of the nutrient is leached in those first drops of water out the bottom. This not only wastes money, it robs the plug of needed fertilizer.

Trays that are seeded and watered-in usually are moved into the proper germination area within two hours, again to help keep moisture constant. If they have to be left out overnight, they're watered by hand again to the right depth before heading into germination.

Record keeping

With what can amount to several thousands trays in daily production for many customers, record keeping is paramount. Throughout Stage 0 production, Raker's Start Team members record every Stage 0 production detail. Media samples are stored for future reference. Seed and tray selections must match. Orders must be sown to the right specifications. The seed lot must be checked against a report to ensure the right lot is used.

The team keeps three sets of records: a printed computer report, a written inventory book, and a log of what trays are sown, when, and with which seed lot. This may seem like a lot of redundant work, but Jenny says these checks and balances help prevent problems at the outset and provide an important paper trail for later problem solving.

"So many things can go wrong in plug production," Jenny says. "Starting things out right helps contribute to a uniform plug. Starting with good seed and contributions on every level—especially in Stage 0—are important. There are few excuses not to make a uniform plug."

October 1997.

Heating up the Pansy Plug Market

by Sherri Bruhn

Nestled in the mountains of northern Georgia, 400,000 square foot of glass blinks in the summer sunlight as the morning mist begins to roll away. It's almost 8 a.m., and Speedling's employees have already been hard at work for a couple hours. They're hustling to beat the midday heat that can bring temperatures to a scorching 95°F inside, even with the most modern environmental controls. They expect no mercy to quench July's relentless heat wave; there are demands to meet and inventories to fill. And the Blairsville location will hit those deadlines, maintaining Speedling's signature for quality that denies the heat was even an issue.

Formerly a Greiling Farms facility, the Blairsville greenhouse has come a long way since Speedling purchased it in 1997. Among nine Speedling facilities across the United States, Blairsville is an impressive earner. Blairsville contributed to 40% of Speedling's total ornamental sales. A large share of those earnings can be attributed to pansy plug production—over 30,000 trays alone were produced in the last week of July. The expertise and quality is a tribute to Greiling Farms' reputation as a top producer; the staying power is a result of Speedling's automation philosophy and expanding customer base.

High Numbers, High Efficiency

The weekly pansy numbers are mind-boggling: They average 27,000 trays each week. Estimate that each tray holds 400 cells, and that equals roughly 11 million pansy plugs per week. During the peak season, pansy production becomes a matter of habit, something as natural as blinking. And the benches turn quickly; Dutch tables move out of the germination chamber and into their respective zones. At the busiest moments, there's constant motion as each table moves down the line of its development, disengaging from one row and clicking into the next. A healthy sea of green stretches acre upon acre, beneath a massive Venlo structure that rises to twenty feet at the peak. It's so seemingly efficient that it almost feels empty inside.

"The design of the greenhouse is so efficient that you can take a lot of advantage of it," remarks head grower Regina Coronado. "You don't need people to pull shades because you have a computer that reads the light levels, and [the curtains] close automatically. We don't have to worry about pulling weeds; there are no weeds in here. And, of course, this rolling method makes it easier for everybody."

The pansies begin their voyage in the germination chamber. From raw seed, it takes about four days to germinate at a maintained temperature of 62°F, with humidity levels between 88 and 90%. With Ball seed, it takes about two days, and Novartis Seed's PreNova seeds will germinate overnight. Once seeded, automated systems take over, and the plants aren't touched until they're shipped out.

Ken Ashworth, traffic manager, says racking and boxing up plants for shipment is an eight-hour task at the busiest times. But loading and shipping don't require a separate crew. Labor tasks are rotated, allowing everyone to get involved in each part of the process, from beginning to end. Many former Greiling customers in the Southeast have remained faithful destinations for Speedling's plugs, but the customer base has expanded to require shipments across the nation and even to some international clients.

"The customer base is the same; we're serving the same customer as we were before," says Ken. "Under Greiling, this facility served mainly the

Southeast because they had so many facilities, and each facility was regionalized. What we've seen now is more shipments outside the Southeast. We're doing a lot more air freight than we ever did before."

Quality Pansies, Quality People

The cool mornings come and go all too quickly, but these hot summers are routine for those who call Georgia home. And the pansies seem to adapt too, thanks to a knowledgeable staff that treats them the right way. Regina was transferred to Blairsville from Speedling's Texas location, and her staff believes she has made worthwhile changes to the growing system.

"The hot, dry spells make it hard to get the chemicals on," explains grower Billy Roberson. "Then Regina came in with a lot of experience. And she grew where it was terribly hot, so she could show us different things they did there that helped them out. She's helped us out quite a bit."

"We have a supervisor who lets us put our knowledge to work," adds grower Jim Herrison. "Gradually, as you learn about the plants, you learn they can do with less."

Regina's guidance has led to fewer chemical sprays. She also understands the nuances of light levels and makes sure the plants don't dry out. And a nitrate-based fertilizer and the growth regulator A-Rest have worked with the high temperatures to resist stem elongation. Hesitant to accept high praise, Regina credits her staff with making the difference.

"We have some qualified people, and I think part of my success is that of the growers' success," she says. "They're the ones who do 80% of what's going on in the greenhouse. Each grower is the manager of his or her own zone, and they know their zones better than anyone else. They're responsible for what happens."

Accountability, reliability, and teamwork seem to be the underlying principles behind the praise. It goes hand in hand with the state-of-the-art Priva computers and the open communication that exists among staff members. The staff, in turn, seems to appreciate a flexible work environment. In an area with a 2.3% unemployment rate, it becomes exceedingly important to keep everyone satisfied and on board.

"We've got a wonderful group of people," says division manager Bill Powell. "There's not one person out there I'd like to see leave. No one's dragging the group down; they've all got a good work ethic. It's like a family."

Family values such as these are important when looking at a place that will put out 70 to 80 million pansy plugs this year. It's vital that everyone's

schedule fits together to ease the burden labor shortages make during the challenging peak seasons.

Moving Forward

And the challenges will persist, according to marketing manager Barry Ruta. He projects that production numbers will continue to soar. While Blairsville's acquisition served the pansy purpose that couldn't be met in other facilities, now it's been taxed to its limits.

"Blairsville was purchased as an augmentation to our Florida location. Because of the climate [in Florida], we couldn't get involved with the pansy season and we wanted a tie-in to that market," says Barry. "The pansy program has been considerably more than we anticipated. We're bursting at the seams at the moment."

But exploding pansies aren't the only things on the menu. Bill expects to move into cut flower and perennial production as well. Cut flower varieties will include lisianthus, larkspur, and snapdragons. Bedding plant production has also included other varieties. Part of this is due to the successful Pansy Day held for the first time last season. This year's trials will embrace other cool-weather crops. He says that he feels Speedling is capable of doing a good job with all these new additions, and it serves the dual purpose of providing finished crops to customers outside the Southern region.

"We have the location and the technology," Bill says. "People have been asking for other kinds of plugs, and they fit into the spaces. We can set them up any way we want. [We'll also provide] a turn of finished products sold under contract for the Midwest so we're not competing with our Southern customers."

Customers haven't complained thus far. In fact, the numbers are rising to include new customers who had formerly been clients of Florida's Natural Beauty and Greiling Farms. Other customers are likely to swarm as a result of increasing coordination with the Florida Speedling location. Bill says the shift has gone South to meet the demands for pansies and cool-weather plants not available elsewhere.

Some things have been sacrificed: Poinsettias have dropped off completely. But that loss isn't making much of a difference. "We are gaining market share each year but also seeing increases within our existing customers," says Barry. "We're going to need to expand further, either through additional acquisitions or expanded locations."

In other words, the temperature just got a little bit hotter.

September 1999.

Index

Page numbers for photographs are **bold**, and page numbers for tables are *italic*.

A

A-Rest, 44–47
ageratum, 50, 125
algae, controlling, 152
alkalinity of irrigation water, 11–12
alyssum, 125
aster, 60
average daily temperature (ADT). *see also* temperature
  and finishing time, 54–55
  and root-to-shoot ratio, 81, 83–84

B

Ball Vigor Index (BVI), 2
bar coding, 107, 173–78, **175**
begonia, Non-stop, 69–70
begonia, fibrous
  culture, 125–26, **151**, 151, **152**, 152–55, **155**
  flowering, 50
  germination, 58
  nutrient deficiency, 89–91
Blackmore Company
  seeding equipment, 114, 117–18
  tray size standards, 121
Bob's Market and Greenhouse (WV), 2–3, 5
Bonzi
  applied to seed, 48–49, *49*
  and root-to-shoot ratio, 82
  boron defiency, 70–71
Bouldin & Lawson
  seeding equipment, 114, 118
  tray size standards, 121–22
  Brodbeck Greenhouse (OH), 131–34

C
C. Raker & Sons (MI)
  bar coding, 107
  bench-top germination, **87**
  customer service, 4–6
  grower profiles, 161–70, **162**, 179–84, **182**
  replugging equipment, 112–13
  seeding equipment, 105–6, 114, 117
calcium, 16–17
celosia
  A-Rest effect on, 47
  culture, 126
  flowering, 50, 72, **77**, 78
  germination, 60
  chemical growth regulators. *see* plant growth regulators (PGRs)
Cherry Creek Systems, 111
coleus, 126
cropping time, reducing, 49–55
customer service, 4–5

D
dahlia
  A-Rest effect on, 47
  culture, 126, **126**
  germination, 60
damping-off disease, 72–74
dianthus
  A-Rest effect on, 47
  culture, 127
  germination, 59
DIF
  and finishing time, 54
  and root-to-shoot ratio, 81, 83
dusty miller
  culture, 127
  germination, 59

E
electronic data interchange (EDI), 5–6, 168

F
fertilizer. *see* nutrition
finishing time, reducing, 49–55
First Step Greenhouses (CA), 114, 117
Flier USA
   replugging equipment, 107, 111
   seeding equipment, 114, 117–18
   tray size standards, 121, 123
flower initiation, 50–51, 71–72, 77–78
fungi, damping-off, 72–74
fungicides, 69, 73
fungus gnats, 74, 91–94

G
geranium
   Brodbeck Greenhouse (OH), 131–34
   culture, 127, **131**, 134–36, **135**
   germination, 59
   quality control of plugs, **75**
gerbera, 127
germination. *see also specific crops*
   defined, 94
   environment and, 56–66, 96–98
   in germination chambers, 60–66, *61*
   in greenhouse, 86–88
   phases, 94–96
   seed testing, 30–31
Green Circle Growers (OH), 161–70
greenhouse automation, 170–73, **171**

H
Headstart Nursery (CA), 105

Need more information about plug production?

INCLUDED: three separate 20 in. x 28 in. wall-size charts with crop-by-crop growing information.

Plug & Transplant Production — A Grower's Guide

by Roger C. Styer, Ph.D. and David S. Koranski, Ph.D.

The first complete reference for beginning to experienced growers — everything you need to know about growing plugs and transplants. This 400-page hardcover manual includes: selecting structures to production systems, understanding seed physiology to scheduling plugs, 16-page four-color section for disease diagnosis.

$84.95* Item #B031   ISBN: 1-883052-14-9   400 pages, hardcover, ©1997

GrowerVision on Plug Production

This six-tape video series features Dr. Roger Styer, coauthor of *Plug and Transplant Production: A Grower's Guide*. The series is shot entirely in plug greenhouses around the country and will take you through the "tricks of the trade." Covering topics such as Stage 0, germination, nutrition, growth regulators, keys to growing major crops, and more, you will see and hear how to consistently grow quality plugs. A workbook containing outlines of each tape along with additional information accompanies the series. Use these tapes to train new growers, review cultural practices before the season starts, or solve problems. No plug grower should be without this innovative, information-packed video series!

Great for Training New Employees!

$149.95*   (Price includes six video tapes and workbook.)
Item #VID7   Tapes come in NTSC format. International customers can request
PAL or SECAM format for $239.95 — use item #VDP7.

* Prices *do not* include shipping and handling charges.
  Please call for your final total before sending payment.

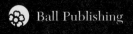 Ball Publishing

Phone: 1-888-888-0013 / 1-630-208-9080 (outside U.S.)
Fax: 1-888-888-0014 / 1-630-208-9350 (outside U.S.)
Internet: www.ballbookshelf.com
Mail: P.O. Box 9, Batavia, IL 60510 U.S.A.

Log On!

- Books
- Video
- Software

www. ballbookshelf .com

- Your online resource for floriculture books, video, and software

- Over 100 titles available on topics such as bedding plants, perennials, pests and diseases, cut flowers, marketing, and more!

- Completely secure

- Full e-commerce capability with email confirmation

- Fully searchable by subject, author, and title

- Accepts Visa®, MasterCard®, and American Express®

Ball Publishing

STRATHCLYDE UNIVERSITY LIBRARY

30125 00373406

KV-575-458

DUE
- 9 JUN 2006

DUE
- 1 MAR 2010

DUE
- 9 JAN 2007

DUE
1 2 APR 2010

DUE
1 5 SEP 2010

ANDERSONIAN LIBRARY
WITHDRAWN
FROM
LIBRARY
STOCK
UNIVERSITY OF STRATHCLYDE

Dr David Chappell RIBA MA PhD currently lectures in construction, building law and contractual procedures. He has previously worked as an architect in public- and private-sector practice and has experience as contracts administrator for a building contractor. He is author of *Contractual Correspondence for Architects, Report Writing for Architects,* and *Contractor's Claims.* He is also joint author with Dr Vincent Powell-Smith of *Building Contract Dictionary* and *JCT Intermediate Form of Contract: an Architect's Guide.*

Dr Vincent Powell-Smith LLM DLitt FCIArb, Commander of the Order of Polonia Restituta, sometime Lecturer in Law at the University of Aston Management Centre, now acts as a consultant specialising in building contracts and as a practising arbitrator. A well known conference speaker, he has been Legal Correspondent of *Contract Journal* for the past twelve years and is a regular contributor to *The Architects' Journal* and *International Construction.* A former member of the Council of the Chartered Institute of Arbitrators, for several years he was a member of the Minister's Joint Advisory Committee on Health and Safety in the Construction Industry. He has written a number of highly successful titles on construction law topics and is joint editor of *Construction Law Reports.*

Architectural Press Legal Guides

JCT Minor Works Form of Contract

#18561943

David Chappell and Vincent Powell-Smith

An Architect's Guide to the Agreement for Minor Building Works

UNIVERSITY OF
STRATHCLYDE LIBRARIES

The Architectural Press: London

First published in 1986 by The Architectural Press Ltd, 9 Queen Anne's Gate, London SW1H 9BY

© D M Chappell and V Powell-Smith 1986

All rights reserved. No part of this book may be reproduced, stored in a retrieval system or transmitted in any form or by any means, electronic, mechanical, photocopying or otherwise, for publication purposes, without the prior permission of the publisher. Such permission, if granted, is subject to a fee, depending on the nature of the use.

The only exceptions to the above prohibition are the sample letters provided in this book as models to be used by readers in architectural practice. These letters may be freely reproduced, on the clear understanding that the publishers and the authors do not guarantee the suitability of any particular letter for any particular situation. Legal advice should be taken in case of doubt.

British Library Cataloguing in Publication Data
Chappell, David
  JCT minor works form of contract: an architect's guide to the agreement for minor building works
  1. Joint Contracts Tribunal. Agreement for minor building works. 1980. 2. Building—Contracts and specifications—Great Britain
  I. Title II. Powell-Smith, Vincent
  692'.8 TH425

ISBN 0-85139-886-3

Typeset by Phoenix Photosetting, Chatham
Printed and bound in Great Britain by
Biddles Ltd, Guildford and King's Lynn

D
692·80941
CHA

UNIVERSITY OF STRATHCLYDE
21 SEP 1989
UNIVERSITY LIBRARY

Contents

Chapter 4 Architect

Chapter 5 Contractor

Chapter 6 Employer

Chapter 7 Quantity Surveyor and Clerk of Works

Preface

Ever since the Agreement for Minor Building Works was first published by the Joint Contracts Tribunal in 1968, it has proved to be a very popular form of contract. It has been used consistently, not only for the simple short contracts of moderate price for which it was intended, but also for much larger projects for which it is often not suited at all. It is a very short form, only seven pages in length. Its brevity no doubt accounts for much of its popularity.

Its simplicity, however, is deceptive. That which is omitted may be as significant as the express terms. Like all forms of contract, its terms must be read against a considerable body of case law.

We have addressed this book to the architect, but it will be essential reading for contractors also, because it explains the practical applications of the form.

This book is not a clause-by-clause interpretation. We consider it more useful to tackle the form under a series of topics, such as the architect, employer, contractor, claims, payment and determination, etc. We have eschewed the use of legal language, prefering to express legal concepts in simple terms supported by flow charts, tables and sample letters. Where possible, we have attempted to provide a clear statement of the legal position in a variety of common circumstances together with references to decided cases so that those interested can read further. In short, this book is a practical working tool for all those involved in contracts using the Minor Works Form.

We must express our thanks to Maritz Vandenberg, Director in charge of Book Publishing at The Architectural Press Ltd, for his usual patience, encouragement and diplomacy.

The copyright in MW 80 and in the supporting practice notes is vested in RIBA Publications Ltd, and we are indebted to them for permission to quote from their publications for the purposes of this book.

David Chappell, Vincent Powell-Smith

Chapter 1

The Purpose and Use of MW 80

1.1   The background

The JCT Agreement for Minor Works was first published in 1968; it was revised in 1977. The headnote explained that it was intended for minor building works or maintenance work, based on a specification or a specification and drawings, to be carried out for a lump sum. It was inappropriate for use with bills of quantities or a schedule of rates.

Despite its shortcomings, it was widely used for small projects, and even for larger ones, its main attraction being its brevity and apparent simplicity. The evidence suggests that architects were becoming increasingly dissatisfied with the length and complexity of the main JCT standard form contract (JCT 63) and then – as now – wished to use simple contract conditions wherever possible.

The Minor Works form was extensively revised by a JCT working party in 1979 and a new edition was published in January 1980, and reprinted with corrections in October 1981. In effect, MW 80 is a completely new set of contract conditions. The headnote to the form as issued in 1980 set out its purpose:

'The Form of Agreement and Conditions is designed for use where minor building works are to be carried out for an agreed lump sum and where an Architect or Supervising Officer has been appointed on behalf of the Employer. The Form is not for use for works for which bills of quantities have been prepared, or where the Employer wishes to nominate subcontractors or suppliers, or where the duration is such that full labour and materials fluctuations provisions are required; nor for works of a complex nature or which involve complex services or require more than a short period of time for their execution.'

Not surprisingly, architects found this headnote very misleading, and it was withdrawn in August 1981 and replaced by Practice Note M2, which is much more indicative of the scope of the contract.

1.2   The use of MW 80

Practice Note M2 sets out the criteria for the use of the form. These are:

○   Minor building works to be carried out for an agreed lump sum price under the supervision of an architect.

The lump sum price is what the employer will pay, subject of course to the operation of the clauses dealing with variations, fluctuations and the expenditure of provisional sums. MW 80 can be used as a fixed price contract.

○   The lump sum offer is based on drawings and/or specifications and/or schedules but without detailed measurements.

In principle, the use of bills of quantities is precluded, although it may be that the 'schedules' are in effect scaled down versions of traditional bills.

○   The contract period must be such that full labour and materials fluctuation provisions are not required.

It is doubtful whether this is a critical factor since, in an appropriate case, a suitable fluctuations clause can be incorporated.

○   MW 80 is generally suitable for contracts up to the value of £50 000 at 1981 prices – say £75 000 at 1986 prices.

Contract value is not, however, necessarily a deciding factor. A key factor seems to be the complexity of the works.

○   Where the employer (through the architect) wishes to control the selection of sub-contractors for specialist work the use of the form is *prima facie* precluded.

The practice note points out that this may be done by naming a firm in the tender documents or in a provisional sum instruction, but 'there are no provisions in the form which deal with the consequences . . . nor is there any standard form of sub-contract which would be applicable to such selected sub-contractors'.

Clearly the use of nominated or named sub-contractors as such would require substantial amendments to the form as printed. It may also be noted that there is no currently available standard form of sub-contract for use with ordinary ('domestic') sub-contractors although MW 80 envisages that the contractor may sub-let with your consent.

The practice note suggests that the control of selection of specialists can be achieved by the employer contracting directly with the specialist.

In fact, MW 80 is highly suitable for use on projects which have a simple work content and where bills of quantities are not required. As with IFC 84 – the next contract up in the JCT range – three things debar the use of MW 80:

○   Complexity of work
○   The wish or need for nominated sub-contractors and/or nominated suppliers.
○   The wish to make the contractor (or any sub-contractor) wholly or partly responsible for design.

You should not use MW 80 merely on account of its apparent simplicity or because, sensibly, you dislike the complex administrative procedures of JCT 80 and the comparatively uncharted sea of IFC 84. The brevity and simplicity of MW 80 is more apparent than real because its operation depends to a large extent on the gaps in it being filled by the general law. Some of the more obvious gaps can be plugged by drafting (or getting a lawyer to draft) suitable clauses. To take but one example: nowhere in MW 80 is there any provision dealing with contractor's 'direct loss and/or expense' claims as found in JCT 80 and IFC 84.

This does not mean that the contractor must allow for the possibility of claims in his tender price or that, in appropriate circumstances, he cannot recover them. It merely means that there is no contractual right to reimbursement and that you have no power under the contract to deal with them. As explained in Chapter 10, the contractor can pursue such claims in arbitration or by means of legal action. And in Chapter 10 we make appropriate suggestions for dealing with this familiar construction industry problem.

1.3 Arrangement and contents of MW 80

MW 80 consists of only eight contract conditions, sub-divided in what is not always the most logical way. One of those conditions (Clause 8.0) refers to a separate Supplementary Memorandum which contains the provisions for tax fluctuations, value added tax and statutory tax deductions. The form consists of eight pages, including the back sheet which makes it clear that the form is not for use in Scotland which has its own legal system and applicable rules of law. The actual conditions of contract occupy four pages of the document and are printed in double-column format.
The whole document is arranged as follows:

MEMORANDUM OF AGREEMENT
RECITALS
ARTICLES OF AGREEMENT
Table of Contents
CONDITIONS
1.0   Intentions of the parties
      Contractor's obligation [1.1]
      Architect's duties [1.2]
2.0   Commencement and completion
      Commencement and completion [2.1]
      Extension of contract period [2.2]
      Damages for non-completion [2.3]
      Completion date [2.4]
      Defects liability [2.5]

3.0   Control of the works
       Assignment [3.1]
       Sub-contracting [3.2]
       Contractor's representative [3.3]
       Exclusion from the works [3.4]
       Architect's instructions [3.5]
       Variations [3.6]
       Provisional sums [3.7]
4.0   Payment
       Correction of inconsistencies [4.1]
       Progress payments and retention [4.2]
       Penultimate certificate [4.3]
       Final certificate [4.4]
       Contribution, levy and tax changes [4.5]
       Fixed price [4.6]
5.0   Statutory obligations
       Statutory obligations, notices, fees and charges [5.1]
       Value added tax [5.2]
       Statutory tax deduction scheme [5.3]
       Fair wages resolution [5.4]
       Prevention of corruption [5.5]
6.0   Injury, damage and insurance
       Injury to or death of persons [6.1]
       Damage to property [6.2]
       Insurance of the works – Fire, etc. – new works [6.3A]
                                 – Existing structures [6.3B]
       Evidence of insurance [6.4]
7.0   Determination
       Determination by employer [7.1]
       Determination by contractor [7.2]
8.0   Supplementary memorandum
       Meaning of references in 4.5, 5.2 and 5.3

There is, happily, little supporting documentation which you require. It
consists only of a Supplementary Memorandum (see above) and two
practice notes (M1 and M2). Practice Note M1 merely summarises the
changes from the earlier edition of the form, while M2 has been summarised in section 1.2.

1.4   Contractual formalities

Like any other contract, a building contract is made by an offer, acceptance and consideration. The contractor's tender is the offer, and a
properly worded letter of acceptance from the employer (or from you on
his behalf) will create a binding·contract. Acceptance of the contractor's

tender must be unqualified. An acceptance subject to further agreement does not result in a contract.

It is sensible for a formal contract to be executed by the parties, and it is best to use the printed form for this purpose. In the private sector the preparation of the contract documentation is your responsibility, and you need to take great care.

You will need two copies of the printed form, which you must complete identically, filling in the blanks and making the necessary alterations and deletions. The printed form as completed will then be signed or sealed by the employer and the contractor.

THE ARTICLES OF AGREEMENT

Page 1 should be completed with the descriptions and addresses of the employer and the contractor. The date will not be inserted until the form is signed or sealed.

The first recital is vital because it is here that a description of 'the Works' must be inserted, and that description is important when considering the question of variations, and for other purposes as well. Only two and a half lines are allowed for this description, so be both precise and succinct. Your name (or the name of your firm) must also be filled in. The question of contract documents is discussed in Chapter 3 and the necessary deletions should be made both here, and in the second recital.

This deals with pricing and makes provision for the contractor to price the specification, schedules or a schedule of rates, and it should be appropriately completed. If you have chosen to go out to tender merely on the basis of drawings and the contract conditions only and the contractor also submits a schedule of rates, you would be very unwise to let the reference to the schedule of rates stand. You did not require it and you should not acknowledge its existence.

The third recital is declaratory, and the contract drawings must be signed by or on behalf of the parties, using a simple formula such as:

'This is one of the Contract Drawings referred to in the Agreement made on [date] between [the employer] and [the contractor].'

If a quantity surveyor has been appointed, his name or the name of his firm must be inserted in the fourth recital. If not, the recital should be deleted.

It is also necessary to complete Articles 2 and 3.

Article 2 sets out the contract sum and this should be written out in full with the figures in brackets. Of course, if fluctuations are applicable, this sum will be subject to the provisions of the supplementary memorandum.

Article 3 defines the term 'architect/supervising officer' and you must insert your name and address (or that of your firm) in the blank space even though you are also mentioned in the first recital. MW 80 provides for a designated architect and the contract is in fact inoperable without an architect who is charged with the performance of many important duties

Fig 1.1

Attestation clause for contract under seal

```
SIGNED, SEALED & DELIVERED
by the above-named Employer
in the presence of:
```
}   (**L.S.**)

```
THE COMMON SEAL of the
above-named Contractor was
hereunto affixed in the
presence of:
```
}   (**L.S.**)

```
Company Secretary
```

Table 1.2

Filling in the form

Item or Clause	Comment
Memorandum	Insert names and addresses of the parties. Date to be inserted when contract is executed.
Recital 1	The description of the works must be sufficient to identify them clearly. Contract drawing numbers must be filled in. Delete inappropriate documents.
Recital 2	Delete as appropriate.
Recital 3	Make sure contract documents are signed.
Recital 4	Insert QS's name and address or delete.
Article 2	Insert contract sum in words and figures.
Article 3	Insert your name and address (or your firm's).
Article 4	Delete RIBA or RICS.
Attestation	Decide signature or seal. If seal chosen, provide attestation clause and delete printed version.
Clause 2.1	Insert dates of commencement and completion. Do not put 'to be agreed'.
Clause 2.3	Insert amount of liquidated damages.
Clause 2.5	The usual defects liability is three months. If a different period is required, it should be written in and the 'three months' should be deleted.
Clauses 3.6 and 4.1	The appropriate deletions should be made consistent with the recitals.
Clauses 4.2/4.3	Complete as appropriate to retention percentage chosen. The 5% is usual.
Clause 4.4	If a longer or shorter period is desired, 'three months' should be deleted and another period inserted.
Clause 4.5	This must be deleted if fluctuations do not apply.
Clause 5.4	The fair wages clause must be deleted in private sector work.
Clause 5.5	Delete this unless the employer is a local authority.
Clauses 6.3A and 6.3B	One or other alternative must be deleted and if 6.3A applies the percentage should be written in.

under the contract terms, as discussed in Chapter 4. Indeed, so important is your role, that the employer *must* appoint a successor architect if you retire, resign your appointment, or die. This was laid down in *Croudace Ltd v London Borough of Lambeth* (1986) 6 ConLR 72.

In *Article 4* you must delete one or other of the appointing bodies; which one you delete is a matter for the employer, acting on your advice.

All deletions or alterations in the recitals, articles of agreement and, indeed, in the contract conditions must be struck through and initialled by the employer and the contractor.

Attestation

For some reason best known to the Joint Contracts Tribunal, there is only one printed form of attestation, and that provides for the contract to be executed by signature only. There is, however, a blank space on Page 3 if it has been decided to execute the contract under seal. This is a decision which the employer will have made – on your advice – at pre-tender stage, and there is an important practical difference.

The Limitation Act 1980 specifies a limitation period – the time within which an action may be commenced – of six years where the contract is merely signed by the parties ('a simple contract') or of twelve years where the contract is made under seal ('a speciality contract'). The longer period is beneficial from the employer's point of view and that is why most local and public authorities insist on building contracts being sealed.

Fig 1.1 is the usual form of attestation clause for a contract under seal.

Red wafer seals (obtainable from law stationers) should be used if possible, although it is sufficient to write 'L.S.' (meaning 'in the place of the seal') if they are not obtainable. Most building contractors operate as limited companies and, by law, they are required to have a company seal. Stamp duty is no longer payable on building contracts under seal unless they are complicated by certain matters such as conveyances or leaseback, which will not normally be the case with MW 80.

Whichever method is used – signature or sealing – it is your duty to check over the contract as signed or sealed and make sure that all the formalities are in order. All too often these administrative chores are pushed on one side and then when something goes wrong the legal profession has a field day.

Table 1.2 tabulates the decisions which have to be made on the contract clauses.

All necessary decisions as to the applicable clauses and amendments must be made at pre-contract stage, and you will have notified the contractor accordingly. He is entitled to know the terms on which he is expected to contract. In completing the printed form, you must ensure that the final version for signature or sealing accords with the terms on which the contractor submitted his tender.

Chapter 2

Contract Comparisons

2.1 Introduction

Advising the employer on which form of contract is best suited to the particular project is one of your important functions. This is deducible from Part I of the *Architect's Appointment*, where – under the heading 'Work stage J: Project planning' – paragraph 1.20 says:

'Advise the client on the appointment of the contractor and on the responsibilities of the client, contractor and architect under the terms of the building contract; where required prepare the building contract and arrange for it to be signed by the client and the contractor . . .'

In a certain sense, this puts the cart before the horse, because in the classic statement of the architect's duties – for breach of which you may be sued – the sixth duty is put in this way:

'To consult with and advise the employer as to the form of contract to be used (including whether or not to use bills of quantities) and as to the necessity or otherwise of employing a quantity surveyor . . .'

This sentence – taken from Hudson's *Building Contracts*, 10th edition, page 103, should be engraved on your heart, and advising the employer on which form of contract to use is one of the most neglected of your professional functions. A substantial body of case law has built up as the result of using standard contract forms for purposes for which they were not intended and you should certainly explain to your client the main provisions of whichever form of contract you recommend.

As regards the JCT contracts, JCT has issued a very generalised practice note (Practice Note 20), the most useful feature of which is the comparison which it draws between the range of JCT contracts currently available.

In the private and local authority sectors, the effective choice for work which has been full-designed by the employer's consultants is between one of four Joint Contracts Tribunal forms and the more recent (and innovative) form produced by the Association of Consultant Architects.

The ACA Form – a second and vastly improved edition of which was published in 1984 – is not in our view suitable for truly 'minor works', but – to use JCT terminology – there will be many instances where 'minor works' shade into 'intermediate works'. The essential point about MW 80 is that the contract conditions provide only a skeleton; but certainly where simplicity of administration is desired, the works are uncomplicated, there is no desire for nominated specialists and bills of quantities are not required, MW 80 should be considered.

In Chapter 1 of our book *Building Contracts Tabulated and Compared* (The Architectural Press Ltd, 1986), we have gone into the question of contract choice and comparison in some detail, but for present purposes the critical factors in contract choice are:

o   The scope and nature of the work.

o   The presence or absence of bills of quantities.

o   Lump sum or approximate price.

In the JCT family – where you have decided that bills of quantities are inappropriate – the effective choice is between the 'without quantities' version of JCT 80, IFC 84 and MW 80. The Appendix to JCT Practice Note 20 – the full title of which is 'Deciding on the appropriate form of JCT main contract' – sets out the main differences between these three forms on what it describes as 'matters of major significance in building contracts'. The features which it allocates to MW 80 are:

o   *Date of possession*

MW 80 contains a simple provision (Clause 2.1) stating that the works 'may be commenced' on a specified date. In our view, this particular wording is of no real significance in light of the common law position.

o   *Extension of time*

In contrast to JCT 80 and IFC 84, MW 80 has very brief provisions for extensions of time. The contractor must notify you of his need for an extension; you are to grant a reasonable extension of the contract time 'for reasons beyond the control of the contractor'.

o   *Liquidated damages*

MW 80 has a traditional liquidated damages clauses. Deduction does not depend on your issuing a certificate of non-completion. The clause states simply that 'the contractor shall pay . . .' the agreed amount.

o   *Payment and retention*

Progress payments to be made by the employer if requested by the contractor. The retention is not expressly stated to be trust money.

o   *Variations and provisional sum work*

You are to value variations on a 'fair and reasonable basis' – using any priced documents.

○   Nominated sub-contractors and suppliers
MW 80 does not allow for nomination of either – which you may well consider to be a plus point! The contract is entirely silent about selection of specialists.
○   Fluctuations
Fluctuations for contributions, levy and tax fluctuations only.
○   Partial possession
There are no provisions for partial possession in MW 80
○   Contractor's money claims
There are no provisions for reimbursing the contractor for 'direct loss and/or expense' – but he can recover under the general law.
○   Disputes settlement
A simple arbitration agreement. It does not give the arbitrator an express power to open up, review and revise your decisions. Arbitration can be opened up at any time.
There are, of course, other significant differences between JCT 80, IFC 84 and MW 80. You cannot assume that you have the same powers as are conferred on you under the more detailed forms – because you have not – but this should not deter your recommending MW 80 in those many cases in which it is the most suitable standard form contract.

2.2   JCT contracts compared

Once you have made the decision to use a JCT form of contract for employer-designed works, on a lump sum basis and without bills of quantities, your choice is narrowed to the three best-known and mostly widely used of the JCT forms.
Table 2.1 enables you to see – it is to be hoped at a glance – the position under MW 80 in comparison with JCT 80 (without quantities) and IFC 84.

Table 2.1

MW 80 Clauses compared with those of JCT 80 and IFC 84

MW 80 clause	Description	JCT 80 clause	IFC clause
1	**Intentions of the parties**		
1.1	Contractor's obligations	2	1.1
—	Quality and quantity of work	14	1.2
—	Priority of contract documents	2.2.1	1.3
4.1	Instructions as to inconsistencies, errors, or omissions	2.3 2.2.2.2	1.4 —
—	Contract bills and SMM	2.2.2	1.5
—	Custody and copies of contract documents	5.1	1.6
1.2	Further drawings and details	5.4	1.7
—	Limits to use of documents	5.7	1.8
1.2	Issue of certificates by architect	5.8	1.9
—	Unfixed materials or goods: passing of property, etc.	16.1	1.10
—	Off-site materials and goods: passing of property, etc.	16.2	1.11
2	**Possession and completion**		
2.1 2.4	Possession and completion dates	23.1	2.1
—	Deferment of possession	—	2.2
2.2	Extension of time	25	2.3
2.2	Events referred to in 2.3	25.4	2.4
—	Further delay or extension of time	25.1	2.5 2.6
—	Certificate of non-completion	24.1	2.6
2.3	Liquidated damages for non-completion	24.2	2.7
—	Repayment of liquidated damages	24.2	2.8
2.4	Practical completion	17.1	2.9

MW 80 clause	Description	JCT 80 clause	IFC clause
2.5	Defects liability	17.2	2.10
3	**Control of the works**		
3.1	Assignment	19.1	3.1
3.2	Sub-contracting	19.2	3.2
—	Named persons as sub-contractors	—	3.3.1
3.3	Contractor's person-in-charge	10	3.4
3.5	Architect's instructions	4.1	3.5
3.6	Variations	13	3.6
3.6 3.7	Valuation of variations and provisional sum work	13.4 13.5	3.7
3.7	Instructions to expand provisional sums	13.3	3.8
—	Levels and setting out	7	3.9
—	Clerk of works	12	3.10
—.	Work not forming part of the contract	29	3.11
—	Instructions as to inspection; tests	8.3	3.12
—	Instructions following failure of work, etc.	—	3.13
—	Instructions as to removal of work, etc.	84	3.14
—	Instructions as to postponement	23.2	3.15
4	**Payment**		
—	Contract sum	14	4.1
4.2	Interim payments	30.1	4.2
4.3	Interim payment on practical completion	—	4.3
4.2	Interest in percentage withheld	30.5	4.4
4.4	Computation of adjusted contract sum	30.6	4.5
4.4	Issue of final certificate	30.8	4.6
4.4	Effect of final certificate	30.9	4.7

MW 80 clause	Description	JCT 80 clause	IFC 84 clause
—	Effect of certificates other than final	30.10	4.8
4.5	Fluctuations	38, 39 & 40	4.9
	Fluctuations; named persons		4.10
—	Disturbance of regular progress	26.1	4.11
—	Loss and/or expense matters	26.2	4.12
5	**Statutory obligations, etc.**		
5.1	Statutory obligations, notices, fees, and charges	6	5.1
5.1	Notice of divergence from statutory requirements	6	5.2
5.1	Extent of contractor's liability for non-compliance	6	5.3
—	Emergency compliance	6	5.4
5.4	Value added tax	15	5.5
5.3	Statutory tax deduction scheme	31	5.6
5.4	Fair wages	19A	5.7
6	**Injury, damages and Insurance**		
6.1 6.2	Injury to persons and property and employer's indemnity	20	6.1
6.1 6.2	Insurance against injury to persons and property	21	6.2
6.3A	Insurance in joint names of employer and contractor (new buildings)	22A	6.3A
—	Sole risk of employer (new buildings)	22B	6.3B
6.3B	Sole risk of employer (existing structure)	22C	6.3C
7	**Determination**		
7.1	Determination by employer	27.1	7.1
7.1	Contractor becoming bankrupt, etc.	27.2	7.2
—	Corruption – determination by employer	27.3	7.3

MW 80 clause	Description	JCT 80 clause	IFC clause
7.1	Consequences of determination by employer	27.4	7.4
7.2	Determination by contractor	28.1	7.5
7.2	Employer becoming bankrupt, etc.	28.1	7.6
7.2	Consequences of determination by contractor	28.2	7.7
—	Determination by employer or contractor	28.1	7.8
—	Consequences of determination by either party	8	7.9
N/A	**Interpretation**		
—	References to clauses, etc.	1.1	8.1
—	Articles to be read as a whole	1.2	8.2
—	Definitions	1.3	8.3
—	The architect/supervising officer	Art 3A	8.4
—	Priced specification or priced schedules of work	N/A	8.5

Chapter 3

Contract Documents and Insurance

3.1 Contract documents

3.1.1 Types and uses

Within its limitations of use (see Chapter 1), MW 80 can be used with a wide variety of supporting documents. Taken together, they are termed the contract documents. In principle, they may consist of any documents agreed between the parties to give legal effect to their intentions.

A number of options are set out in the first recital and they may be conveniently considered as follows:

○ The contract drawings.

○ The contract drawings and the specifications priced by the contractor.

○ The contract drawings and schedules priced by the contractor.

○ The contract drawings, the specifications and the schedules, one of which is priced by the contractor.

One of these options, together with the Agreement and Conditions annexed to the Recitals, forms the contract documents. They must *all* be signed by or on behalf of the parties.

Before embarking on a project, study the options carefully to decide upon the combination which is most suitable for the work. Note that the second recital provides that the contractor must price either the specification or the schedules or provide his own schedule of rates. The contractor's own schedule, however, is not one of the contract documents.

The contract drawings

On very small works, for which of course MW 80 is very well suited, it may be quite acceptable to go to tender merely on the basis of MW 80 and drawings. This system can be very satisfactory, provided that every detail of the information required by the contractor for pricing purposes is included on the drawings. You are not precluded from issuing further information by way of clarification in accordance with clause 1.2 as necessary.

Since there is no specification or schedules, the contractor will be expected to provide his own schedule of rates. If you do not wish him to do so, the second recital must be deleted in its entirety. The significance of the priced document is that it is to be used to value variations, if relevant, under clause 3.6. Reference to the contractor's own schedule of rates is included in this clause even though it is not one of the contract documents. Some commentators advise that the contractor's own schedule of rates should always be excluded, but it is difficult to see the reason for so doing.

The contract drawings and the specification priced by the contractor

In practice, this is a very common way of dealing with small projects. If the specification is to be priced, great care must be exercised in preparing it. This system involves the contractor in taking off his own quantities and the cost of so doing is likely to be reflected in the tender figure. The organisational expertise which is incorporated into the specification will determine how useful the priced document will be in the valuation of variations.

The contract drawings and schedules priced by the contractor

This is a variant of the last system. Since the works must be specified somewhere, it is likely that the specifications element will be incorporated into the schedules. Alternatively, depending on the type of work, it may be feasible to put all the specification notes on the drawings. The schedules will normally be quantified, making this system much easier to price from the contractor's point of view. Indeed, often the schedules are really bills of quantities under another name. The contractor needs to take care, however, that his price is inclusive of everything required to carry out the works. This is true even if some items are missed off the schedules, but shown on the contract drawings. The point can be a difficult one. It is discussed in section 3.1.2: clause 4.1 (correction of inconsistencies).

The contract drawings, the specification and the schedules, one of which is priced by the contractor

In practice, this combination would be used on larger works when the schedules would take the form of bills of quantities. The contractor would then normally price the schedules rather than the specification. Once again, the contractor must take care that he prices for everything the contract requires him to do.

In principle, a schedule of work is always to be preferred over a schedule of rates. The former is capable (or should be capable) of being priced out and added together to arrive at the tender figure. A difficulty arises because it requires a broad measure of agreement on the method of carrying out the works. The contractor will have difficulty in pricing a schedule of work if he considers that a totally new approach will show greater efficiency. Some two-stage method of tendering will probably yield best results in such cases when the contractor can be expected to input his suggestions before the schedules are drawn up.

On the other hand, the figures in a schedule of rates cannot be added together to give the tender figure, and the contractor's own schedule should not be accepted unless he has justified the calculation of the overall sum from the basis of the schedule. To do otherwise would reduce the valuation of variations to a farce. A schedule of rates is most useful where the total content of the work is not precisely known at the outset. MW 80 can be used in this way, with a little adjustment, but it is better to consider some other form such as JCT 80 With Approximate Quantities or the Fixed Fee Form of Prime Cost Contract.

You must insert the numbers of the contract drawings in the space provided in the first recital. On large contracts, when bills of quantities are used, it is usual to designate as contract drawings only those small-scale drawings which show the general scope and nature of the work. Under this contract, however, the situation is very different. The total number of drawings is likely to be relatively small and the contractor will need all of them in order to prepare his tender. Since the contractor's basic obligation (clause 1.1) is to 'carry out and complete the Works in accordance with the Contract Documents' you must be sure that the contract documents taken together do cover the whole of the work. Therefore, the contract drawings must be:

o   The drawings from which the contractor obtained information to submit his tender.

o   Sufficiently detailed so that, when taken together with the specification and/or schedules, they include all workmanship and materials required for the project.

The further information which you must provide under clause 1.2 may consist of drawings, details and schedules. Provided that they are merely clarifying existing information, there is no financial implication. If, however, they show different or additional or less work or materials than that shown in the contract documents, the contractor will be entitled to a variation on the contract sum. None of the 'further information' constitutes a contract document.

Every contract document must be signed and dated by both parties to avoid any later dispute regarding what is or what is not a contract document. In practice, this means signing every drawing and the cover of the specification and/or schedules. It is suggested that you endorse each document: 'This is one of the contract documents referred to in the

Agreement dated . . .' or use some other form of words to the same effect.

3.1.2 Importance and priority

The contract documents provide the only legal evidence of what the parties intended to be the contract between them. They are, therefore, of vital importance. In the case of dispute, the arbitrator or the court will look at the documents in order to discover what was agreed.

Clause 1.1 lays an obligation upon the contractor to carry out the works in accordance with the contract documents. The question often arises: what is the position if the documents are in conflict? Clause 4.1 states that nothing contained in the contract drawings or the specification or the schedules will override, modify or affect in any way whatsoever the application or interpretation of the printed conditions. If, therefore, the specification were to contain a clause purporting to remove the contractor's entitlement to extension of time due to exceptionally inclement weather, it would be ineffective because of this provision, which has the effect of reversing the normal legal rule of interpretation. The effectiveness of a clause worded in this way has been upheld in the courts on many occasions.

Although the contract effectively sorts out priorities as between the printed form and the other contract documents, it gives no further guidance as far as priorities among the other contract documents are concerned. Clause 4.1 simply states that any inconsistency in or between the contract drawings and the contract specification and the schedules (which, if accepted, will include the contractor's own schedule of rates) must be corrected and if such correction results in addition, omission or other change, it must be treated as a variation under clause 3.6.

The particular circumstances of each case will determine exactly how the inconsistency is to be treated.

Two main types of inconsistency are common:

○　　Where workmanship or materials are covered in one of the contract documents, but omitted from the other.

○　　Where workmanship or materials shown in one of the contract documents are in conflict with what is shown in the other.

We will confine our consideration to a contract based on drawings and specification which the contractor has priced. If a schedule is also included, the principle is the same, but the facts may be more complex. The first thing to establish is what the contractor has legally contracted to do. Clause 1.1 merely refers to the 'Contract Documents', but that clearly means the documents noted in the first recital, that is, in this case, the contract drawings, the contract specification and the conditions. If, therefore, the inconsistency is, for example, the fact that a handrail is shown without brackets on the drawings, but brackets are specified in the specification, or vice versa, it will be deemed that the contractor has allowed for the brackets in his price. This is probably the case even if the brackets are not in the specification, which the contractor has priced, but are shown on the drawing. In this example, even if the brackets are not shown or mentioned on either document, it is likely that the contractor must supply brackets (presumably the cheapest he can find to do the job) at no additional cost (*Williams* v *Fitzmaurice* (1858) 3 H & N 844). Much, however, will depend on any general terms which you have included in the specification. An expression such as 'The whole of the materials whether specifically mentioned or otherwise necessary to complete the work must be provided by the contractor' would tend to place the responsibility for omissions squarely on the contractor provided that they could be considered 'necessary' to complete the work. In this case, brackets are obviously necessary to support the handrail.

If the documents are in conflict, the position is rather complicated. Since neither drawings nor specification have priority, it is not clear as to which of them the contractor has had reference in formulating his price. It is tempting to consider that the key document is the specification, since the contractor has priced it. Clause 1.1 clearly indicates that this is a wrong view of the situation. In pricing the specification, the contractor must have regard to the totality of the documents. We take the view that, if the documents are in conflict, it is for you to instruct the contractor as to which document is to be followed in the particular instance, and the contractor is not to be allowed any addition to the contract sum nor is there to be any omission, the contractor being deemed to have included for whichever option you chose. If, however, you solve the problem by omitting the work or changing it to something other than is contained in either of the contract documents, it falls to be valued in accordance with clause 3.6 in the usual way. Although this view may be thought to impose undue hardship on the contractor in certain circumstances, particularly as the inconsistency is due to your oversight, it is the only interpretation that appears to take account of the contract as a whole. The contractor bears a great responsibility to examine the documents thoroughly when pricing. If the contract is on the basis of drawings and priced schedules which are in fact fully developed bills of quantities, the situation remains the same. The employer does not warrant the accuracy of the bills in this instance unless you include a clause to that effect. This, of course, is in complete contrast to JCT 80 and points up one of the great dangers in using this form for works of a greater value than that for which it is intended.

3.1.3 Custody and copies

There is no specific provision in the contract regarding the custody of the contract documents and the issue of copies to the contractor. It is probably sensible to keep the original yourself, but if the employer wishes to have it, be sure you have an exact copy. Although you are not expressly required to do so, it is good practice to make a copy for the contractor and certify to him that it is a true copy. This can be simply done by binding all the documents together and inscribing the certificate on each document including the drawings. It is sufficient to state 'I certify that this is a true copy of the contract document.'

It must be implied in the contract that you provide the contractor with two copies of the contract drawings and specification and/or schedules, otherwise he would be unable to carry out the works. It is usual to provide such copies free of charge.

Any further drawings, details or schedules which you provide under clause 1.2 are not contract documents. They are intended merely to amplify or clarify the information in the contract documents. You are obliged only to supply such drawings as are 'necessary for the proper carrying out of the Works'. It is an implied term of the contract that you will issue such additional information at the correct time: *R. M. Douglas Ltd* v *CED Building Services* (1985) 3 ConLR 124. Failure to do so is a breach on your part for which the employer is responsible. The contractor would have a legitimate claim for an extension of time without the necessity for any prior application for the information. If he suffers disruption, he may also have a claim for damages which he could pursue at common law.

3.1.4 Limits to use

The contract contains no express terms to safeguard your interests in the drawings and specification and there is no express prohibition on the employer from using the contractor's rates and prices for purposes other than this contract.

The general law, however, covers the position. You retain the copyright in your own drawings and specification (unless you expressly relinquish it) and neither the contractor nor the employer may make use of them except for the purpose of the project. Strictly, you may ask the contractor to return all copies to you after the issue of the final certificate, but in practice it is seldom worth the trouble.

The confidentiality of the contractor's rates and prices is safeguarded by the general rule that a party has a duty not to divulge confidential information to third parties. This is especially true when two parties are bound together in a contractual relationship and to divulge the information would clearly cause harm. The contractor's prices are a measure of his ability to tender competitively and secure work. To divulge his rates to a competitor is a serious matter. In practice, it is not easy for the contractor to ensure, for example, that the quantity surveyor does not make use of his prices to assist him to estimate the cost of other contracts, but that is probably of little consequence.

It may be thought prudent to include a clause in the specification, similar to clause 1.8 of IFC 84, to cover the limitations on the use of documents. It does no harm to remind the parties of their obligations in this respect.

3.2 Insurance

3.2.1 Injury to or death of persons

Under clause 6.1 the contractor assumes liability for and indemnifies the employer against any liability arising out of the carrying out of the works in respect of personal injury or death of any person, unless due to act or neglect of the employer or any person for whom he is responsible. The persons for whom the employer is responsible will include anyone employed by him and paid direct such as a directly employed contractor, the clerk of works (if any) and yourself.

In practical terms, the employer will be responsible for the injury or death of any person only if the injury or death was caused by his or his agent's act or neglect. In the event of a claim being made, it will normally be made against the employer. The employer will then join the contractor as a third party in any action and claim an indemnity from him under this clause.

The contractor must maintain and cause any sub-contractor to maintain such insurances as are necessary to cover his liability. No sum of money is stated, nor is space left for a sum to be inserted. The onus is left on the contractor to insure adequately. This would be cold comfort if the employer found that the contractor was under-insured in the event of a claim and had insufficient funds. We suggest that you advise the employer to get the opinion of his insurance broker as to the amount which should be included. Such amount can be stated either on the printed conditions or as a separate clause in the specification. Clause 6.4 gives the employer the right to require evidence that the insurance has been taken out, but there is no provision for the employer to take out the necessary insurance himself and deduct the amount from the contract sum if the contractor defaults. None the less, failure on the part of the contractor to insure is a breach of contract for which the employer could recover damages at common law. In practice, it is certain that the employer would set-off such sums against money payable to the contractor.

The requirement that the contractor insures is stated to be without prejudice to his liability to indemnify the employer. In the case of an insurance company failing to pay in the event of an accident, the contractor is bound to find the money himself.

The word 'maintain' is used deliberately in respect of the contractor's and sub-contractor's insurances to indicate that the requirements of clause 6.1 would be satisfied if the contractor and sub-contractors already have general insurance cover in an adequate amount.

It is always wise for you to advise the employer to retain the services of an insurance broker to give advice about the insurance provisions of the contract. You should get the broker to inspect all the relevant documents and to certify to you that they comply with the contract requirements. It is essential that the insurance be in operation from the time the contractor takes possession of the site.

3.2.2 Damage to property

Under clause 6.2, the contractor assumes liability for, and indemnifies the employer against, any liability arising out of the carrying out of the works in respect of damage to any kind of property provided that it is due to the negligence, omission or default of the contractor or any sub-contractor or any of their respective employees or agents.

The contractor's liability under this clause is limited compared with his liability under clause 6.1. The contractor or sub-contractor must be at fault for the indemnity to operate. The clause is stated to be subject to clause 6.3A/B which provides for insurance against specific named risks.

The contractor must insure and cause any sub-contractor to insure against his liabilities under clause 6.2. Although the contract does not provide for it, you would be wise, on the advice of the employer's broker, to stipulate the amount of cover required. The general comments with regard to inspection of documents in section 3.2.1 are also applicable to this clause.

There is no provision in MW 80 for insurance against claims arising due to damage caused by the carrying out of the works when there is no negligence or default by any party. Most standard forms of contract contain such provision. A provisional sum is included in the contract, if required, to cover specific risks, for example the carrying out of underpinning works to adjoining property. It is not always necessary to take out such insurance and the provision is probably omitted from this contract in view of the minor nature of the works envisaged. The amount of the contract sum, however, is no indication of the possible risk to neighbouring premises and you would be prudent to assess each project on its own merits. There is no reason why the employer should not, and every reason why he should, take out the appropriate insurance himself with the advice of his broker if circumstances appear to warrant it.

3.2.3 Insurance of the works against fire, etc.

Clause 6.3 deals with the insurance of the works against a collection of specific risks. They are as follows:

'fire, lightning, explosion, storm, tempest, flood, bursting or overflowing of water tanks, apparatus or pipes, earthquake, aircraft and other aerial devices or articles dropped therefrom, riot and civil commotion'.

The clause is divided into two parts, one of which is to be deleted. Each part is applicable to a particular situation as follows:

○   6.3A: A new building where the contractor is required to insure.

○   6.3B: Alterations or extensions to existing structures.

3.2.4   A new building where the contractor is required to insure

The provision is not complex and provides for the contractor to insure in the joint names of the employer and himself against loss or damage caused by any of the specified risks. The insurance must cover the full reinstatement value and include:

○   All executed works.

○   All unfixed materials and goods intended for, delivered to, placed on or adjacent to the works and intended for incorporation.

and exclude:

○   Temporary buildings, plant, tools and equipment owned or hired by the contractor or his sub-contractors.

A percentage is to be added to take account of all professional fees.

Great care must be taken in determining the level of insurance cover. Provision must be made for the possibility that the work is virtually complete at the time of the damage and allowance made for clearing away before rebuilding. Although it is the contractor's responsibility, you must satisfy yourself that the level of cover is adequate and if you are not so satisfied, you must notify the employer and the contractor immediately (Figs 3.1 and 3.2). Despite the absence of express provision, the employer must insure himself if the contractor defaults or fails to insure adequately.

The insurance must be kept in force until the date of issue of the certificate of practical completion. This is the case even if practical completion is delayed beyond the date for completion in the contract. If the contractor already maintains a policy which provides the same type and degree of cover required under clause 6.3A, it will serve to discharge the contractor's obligations. It must, however, be endorsed to show the employer's interest and the contractor must produce documentary evidence at the employer's request.

If damage occurs, the contractor need do nothing until the insurers have accepted the claim. Once accepted, however, he must 'with due diligence' restore or replace work or materials or goods damaged, dispose of débris before proceeding to carry out and complete the works as before. He is entitled to be paid all the money received from insurance less only the percentage to cover professional fees. It is usual for the money to be released in instalments on certificates, but the contract is silent on the point and it

Fig 3.1

Letter from architect to contractor: if he fails to maintain
an adequate level of insurance under clause 6.3A

```
PROJECT TITLE

Dear Sir

I refer to my telephone conversation with your
Mr [insert name] this morning and confirm that you
are not maintaining an adequate level of insurance as
required by clause 6.3A of the Conditions of
Contract.

In view of the importance of the insurance and
without prejudice to your liabilities under clause
6.3A I have advised the employer to arrange to take
out the appropriate insurance on your behalf.  Any
sum or sums payable by him in respect of premiums
will be deducted from any monies due or to become due
to you or will be recovered from you as a debt.

Yours faithfully

Copy: Employer
      Quantity surveyor (if appointed)
```

Fig 3.2

Letter from architect to employer: if contractor fails to maintain an adequate level of insurance under clause 6.3A

```
PROJECT TITLE


Dear Sir


I am not satisfied that the contractor is maintaining
an adequate level of insurance as required by clause
6.3A of the Conditions of Contract.


In view of the importance of the insurance, I advise
you to take out the necessary insurance on the
contractor's behalf without delay.  To this end, I
have already advised your broker of the situation and
you should contact him immediately.  Although the
terms of the contract make no express provision for
you to act on the contractor's default, it is my
opinion that you are entitled to deduct the cost of
the premium from your next payment to the contractor
in this instance.


A copy of my letter to the contractor, dated [insert
date], is enclosed for your information.


Yours faithfully
```

is open to the parties to make any other mutually agreeable arrangement. The contractor is not entitled to any other money in respect of the rein-statement and if there is any element of under insurance, he must bear the difference himself.

3.2.5 Alterations or extensions to existing structures

In the case of an existing building, the employer bears the risk and must insure the works, the existing structures and contents and all unfixed materials as before. The contractor's temporary buildings, etc., are again excluded. There is a proviso that the employer need only insure such contents as are owned by him *and* for which he is responsible. The reason for this is obscure. There is no requirement that the employer insure for the value of professional fees, but they are his responsibility and if he is wise, he will do so. It would be advisable for you to suggest the inclusion of a suitable percentage.

The contractor is entitled to require the employer to produce such evidence as he may need to satisfy himself that the insurance is in force at 'all material times', that is, from the date of possession to the date of practical completion. There is no provision for the contractor to take out insurance himself if the employer defaults, nor to have the amounts of any premiums he pays added to the contract sum. Clearly, however, failure to insure on the part of the employer is a breach of contract, but not one which entitles the contractor to determine his employment in accordance with clause 7.2 (see section 12.3.2) or, it is thought, to treat the contract as repudiated at common law. You must not allow the situation to arise and must remind the employer at the appropriate time if insurance is to be his responsibility.

If any loss or damage does occur, you are to issue instructions regarding the reinstatement and making good in accordance with clause 3.5. If there is any element of under-insurance, this time it is the employer who must make good the difference. The contractor is entitled to be paid for the work he carries out in the normal way.

It must be noted that whether the employer or the contractor is responsible and has to insure, *all* damage to the works by fire, etc., is removed from the contractor's indemnity under clause 6.2.

Unlike the provisions of similar clauses in JCT 80 and IFC 84, there is no provision in this contract for either party to determine the contractor's employment if it is just and equitable to do so after loss or damage. This is presumably a conscious decision on the part of the JCT, and the position is left to be governed by the general law. The total destruction of the works by fire, etc., might bring the contract to an end at common law, but in the majority of cases the parties should be able to come to some mutually acceptable arrangement depending on all the circumstances.

3.3   Summary

Contract documents

○   They may consist of any documents agreed between the parties.
○   Taken together, they must cover the whole of the work.
○   The information provided under clause 1.2 is additional to and not part of the contract documents.
○   They are the only legal evidence of the contract.
○   The printed conditions override anything contained in the drawings or specifications.
○   The correction of inconsistencies may not result in a variation.
○   You have an implied duty to supply the contractor with sufficient copies of the contract drawings and specification and/or schedules for him to carry out and complete the work.
○   Further information necessary must be issued at the correct time or the contractor may claim an extension of time and damages at common law.
○   No drawing or specification must be used for any purpose other than the contract and the contractor's prices must not be divulged to third parties.

Insurance

○   Employer's indemnity covers personal injury and death and damage to property – subject to certain exceptions.
○   Contractor must insure to cover the indemnities.
○   The employer has no right to insure if the contractor defaults.
○   There is no provision for insurance to cover damage to property not the fault of either party.
○   New work must be insured by the contractor against the specified risks.
○   Alterations and extensions and existing work must be insured by the employer against the specified risks.

Chapter 4

Architect

4.1   Authority and duties

The contract contains express provisions regarding the extent of your authority and the duties imposed upon you. These provisions are discussed in detail below (4.2). It would be quite wrong to think, however, that the provisions specifically set out in the contract are the end of the matter. The contract is very brief and, even in the case of a much more comprehensive contract form such as JCT 80, you have obligations which are sensible, but not always immediately apparent.

From the contractor's point of view, your authority is, indeed, defined by the contract. Thus, if you attempt to exceed this authority, the contractor may, quite rightly, ignore you. In fact, if the contractor carries out an instruction which the contract does not authorise you to issue, the employer is under no legal obligation to pay for the results. In such a case the contractor cannot take any legal action against you under the contract, to which you are not a party, but you may find that he has grounds for redress against you in tort for negligence. Of course the negligence has to be proven and his chances of success will depend upon all the circumstances. In order to prove negligence three criteria must be satisfied:

○   There must be a legal duty of care, *and*
○   there must be a breach of that duty, *and*
○   the breach must have caused damage.

To take an example, if you instructed the contractor to carry out work on land owned by the employer, but outside the site boundaries as shown on the contract drawings, you would be acting outside any express or implied authority of the contract. The contractor would be foolish to carry out such an instruction because he would leave himself open to action by the employer for trespass at the very least. You would possibly find yourself on the receiving end of actions from both contractor (in tort) and employer (under your contract with him).

Table 4.1
Architect's powers under MW 80

Clause	Power	Comment
3.2	Consent in writing to sub-contracting	Consent must not be unreasonably withheld
3.4	Issue instructions requiring the exclusion from the Works of any person employed thereon	The power must not be exercised unreasonably or vexatiously
3.5	Issue written instructions	
3.5	Require the contractor to comply with an instruction by serving written notice on him	The contractor has 7 days from receipt of the written notice in which to comply. If he fails to do so the employer may employ and pay others to carry out the work and may recover the costs involved from the contractor
3.6	Order an addition to or omission from or other change in the Works or the order or period in which they are to be carried out	The variation is to be valued by the architect on a fair and reasonable basis unless its value is agreed with the contractor before the instruction is carried out
3.6	Agree the price of variations with the contractor prior to the contractor carrying out the instruction	

Your powers and duties (see Tables 4.1 and 4.2) flow directly from your agreement with the employer. It is always prudent to enter into a formal written contract following the terms of the RIBA *Architect's Appointment*. This contract will determine the precise extent of your authority, but it is important to remember that, whatever the terms of your appointment may be, it has no effect upon the building contract between contractor and employer. Therefore, provided that the exercise of your authority is within the limits laid down in the building contract, the contractor may safely carry out your instructions, take notice of your certificates, etc., without worrying whether you have obtained the employer's consent.

For example, the Conditions of Appointment state (clause 3.3) that you must not make any material amendment to the approved design without the employer's consent, and lay an obligation upon you to notify the employer if the total authorised expenditure is likely to be varied (clause 3.4). The contract, however, empowers you (clause 3.6) to issue instructions which may both alter the design and increase the total cost of the works. The contractor must carry out such instructions and the employer is bound to pay. If the employer did not consent to the alteration in design

Table 4.2
Architect's duties under MW 80

Clause	Duty	Comment
1.2	Issue any further information necessary for the proper carrying out of the Works	
1.2	Issue all certificates	See Table 4.7
1.2	Confirm all instructions in writing	Clause 3.5 specifies the procedures
2.2	Make in writing such extension of time as may be reasonable	If it becomes apparent that the Works will not be completed by the stated completion date and the causes of delay are beyond the control of the contractor and the contractor has so notified the architect
2.4	Certify the date when in his opinion the works have reached practical completion	
2.5	Certify the date when the contractor has discharged his obligations in respect of defects liability	The contractor's obligation is limited to remedying defects, excessive shrinkages or other faults appearing within the defects liability period and which are due to materials or workmanship not in accordance with the contract *or* to frost occurring before practical completion
3.5	Confirm oral instructions in writing	This must be done within 2 days of *issue*
3.6	Value variation instructions	The valuation is to be on a fair and reasonable basis and where relevant the prices in the priced documents must be used
3.7	Issue instructions as to expenditure of provisional sums	The instruction is to be valued as a variation under Clause 3.6
4.1	Correct inconsistencies in or between the contract drawings, specification and schedules	If the correction results in a change it is to be valued as a variation under Clause 3.6
4.2	Certify progress payments to the contractor	If the contractor so requests. Such payments are to be certified at intervals of not less than 4 weeks calculated from commencement

Table 4.2 *continued*

Clause	Duty	Comment
4.3	Certify payment to the contractor of 97.5% of the total amount to be paid to him	This must be done within 14 days of the date of practical completion certified under Clause 2.4
4.4	Issue a final certificate	Provided the contractor has supplied all documentation reasonably required for the computation of the final sum and the architect has issued his Clause 2.5 certificate (defects liability). The final certificate must be issued within 28 days of receipt of the contractor's documentation
6.3B	Issue instructions for the reinstatement and making good of loss or damage	Should loss or damage be caused by Clause 6.3B events

or the increased expenditure, his remedy lies with you.

If you fail to carry out your duties under the contract, it will generally be held to be a default for which the employer will be liable to the contractor (*Croudace Ltd* v *London Borough of Lambeth* (1986) 6 ConLR 72). Failure to certify progress payments at the right time is an example of such a default, but not failure to include the amount the contractor considers is due (*Killby & Gayford* v *Selincourt Ltd* (1973) 3 BLR 104).

In performing your duties you will be expected to act with the same degree of skill and care as the average competent architect (see *Chapman* v *Walton* [1833] All ER Rep 826). If you profess to have skills or experience which are greater than average, you may be judged accordingly. Thus, if you hold yourself out as, for example, an expert in historic buildings, you will be expected to show a higher degree of skill in that particular area. If something goes wrong, to say that fellow architects without that specialisation would have taken the same action will be insufficient defence.

This is not the place to discuss in detail your general duties to your client and to third parties. One aspect of those duties, however, does affect your administration of the contract. That is your duty to your client to be familiar with those aspects of the law which affect your work. You are not expected to have the detailed knowledge of a specialist contracts lawyer, but you must be capable of advising your client regarding the most suitable contract for a particular project. This implies that you must have, at least, a working knowledge of the main forms of contract. In addition, you must be able to give detailed advice on the contract of your choice. Although problems may arise during the currency of the contract which clearly demand the attentions of a legal expert, your client will be less than impressed and may well consider you to be incompetent if you are unable to explain the basic provisions of the contract and deal with the day-to-day running of the job without legal assistance. If your client is put to unnecessary expense due to the fact that you have inadequate knowledge of the contractual provisions, he may well sue. You are expected to be aware of decisions of the courts relevant to your field of knowledge.

Up to the moment the employer and contractor enter into the contract, you have been acting as agent with limited authority for the employer. During the contract you are expected to continue to act as the employer's agent, but also to administer the terms of the contract fairly between the parties. The employer may find the change difficult to understand and, to avoid problems, it is wise to explain the situation to him, perhaps at the same time as you send the contract documents for his signature (Fig 4.3). In acting between the parties you are not in a quasi-arbitral role and you are not immune from any action for negligence by either party. The point was established by *Sutcliffe* v *Thackrah* [1974] 1 All ER 319. The result is that you may find yourself in the position of deciding whether you are in default yourself, e.g. in considering an extension of time. If the employer suffers loss or damage thereby, he may well seek to recover that loss from you. It might be thought that your position is slightly less hazardous under this contract than it is under JCT 80 or IFC 84 because of the absence of a provision enabling the contractor to claim 'loss and expense' under the contract as a result of your defaults, e.g., through late supply of information. However, if the contractor brought a successful action against the employer at common law, the employer might well be able to recoup his loss from you.

The extent of the authority and duties of an employee of the employer, for example in local or central government, is not always very clear. If you are in this position, the contractor may well consider that you are acting as agent for the employer and that any instruction which you may issue which is not expressly empowered by the contract is, in effect, a direct instruction from the employer. The dangers of incurring unexpected costs are obvious and you should ensure that the extent of your authority is made clear to the contractor at the beginning of the contract (Fig 4.4).

As architect employee, your duty to act fairly between the parties remains, but it is admittedly difficult to convince the contractor that you are acting impartially. The administration of the contract under these circumstances calls for complete integrity, not only on your part, but on the part of the employer who has a duty to ensure that you carry out your duties properly in accordance with the contract (see *Perini Corporation* v *Commonwealth of Australia* (1969) 12 BLR 82). Although it is usual for the actions of local authority employees to be governed by Standing Orders of the Council, they are of no concern to the contractor unless they have been

Fig 4.3
Letter from architect to employer: explaining his duty to
act impartially


```
PROJECT TITLE

Dear Sir

[Insert the main point of the letter and then
continue:]
This is probably an opportune moment to explain the
nature of the additional responsibility which I carry
during the currency of the contract.  Until the
contract is signed I am required, in accordance with
the Conditions of my appointment, to act solely as
your agent within the limits laid down in those
Conditions.  Thereafter, although I continue to act
for you as before, I have the additional duty of
administering the contract conditions fairly between
the parties.  In effect, this means that I must make
any decisions under the contract strictly in
accordance with the terms of the contract.  If you
require any further explanation of the position, I
would be delighted to meet you for that purpose.

Yours faithfully
```

Fig 4.4
Letter from architect to contractor: if architect is employee in local authority, etc.

PROJECT TITLE

Dear Sir

Possession of the site will be given to you by the employer on the [*insert date*] in accordance with the contract provisions.

In order to avoid any misunderstandings which might arise in the future I should make clear that, as far as you are concerned, the extent of my authority is laid down in the contract. Although I am an employee of [*insert name of employing body*], I have no general power of agency to bind the employer outside the express contract terms. If I have cause to write to you on behalf of the employer, I will clearly so state. If any matters fall to be decided by me under the contract, I will make such decisions impartially between the parties.

Yours faithfully

specifically drawn to his attention at the time of tender. Thus an order that only chief officers may sign financial certificates would not avail you as a defence if you were unable to issue a certificate because the chief officer was unavailable. The contractor is entitled to rely on your apparent authority.

The best summary of the legal situation is this: 'An architect is usually and for the most part a specialist exercising his special skills independently of his employer. If he is in breach of his professional duties he may be sued personally. There may, however, be instances where the exercise of his professional duties is sufficiently linked to the conduct and attitude of the employers so as to make them liable for his default.' (Mr Justice Kilner Brown in *Rees & Kirby Ltd* v *Swansea Corporation* (1983) 25 BLR 129). Although this case subsequently went on appeal, it is suggested that Mr Justice Kilner Brown's statement remains an accurate expression of the law.

4.2   Express provisions of the contract

Article 3 of MW 80 provides for the insertion of the name of the architect. The person so named will be the person referred to in the conditions whenever the word 'architect' appears. It is common practice to insert the name of the architectural firm (or the name of the chief architect in a local authority) because of the difficulties which could arise in having to reno-minate every time the project architect left the firm, died or retired. It is seldom the cause of any dispute, but it is prudent to notify all interested parties of the name of the authorised representative, i.e., the project architect, who will administer the contract on a day-to-day basis (Fig 4.5). Changes in the identity of the project architect should be notified in the same way. The case of *Croudace Ltd* v *London Borough of Lambeth* (1984) 1 ConLR 12, establishes that the employer is under a duty to appoint another architect if, for example, the named architect resigns, retires or dies.

It is not strictly necessary for the architect named in Article 3 to sign all letters, notices, certificates and instructions personally, but they must be signed by the registered architect duly authorised 'for and on behalf of . . .' If you are the authorised representative, it is not sufficient for you to sign your name only, even though you may be using headed stationery, because:

○   The letter, certificate, etc., must be signed by or on behalf of the architect named in the contract.

○   Otherwise the letter may be deemed to be written on your own behalf with serious financial consequences.

Do not:

○   Use a rubber stamp.

○   Sign someone else's name and add your initials.

Fig 4.5
Letter from architect to contractor: naming authorised
representatives


```
PROJECT TITLE

Dear Sir

This is to inform you formally that the architect's
authorised representatives for all the purposes of
the contract are:

[insert name] - Partner in charge of the contract.
[insert name] - Project architect.

Until further notice the above are the only people
authorised to act in connection with the contract.

Yours faithfully



for and on behalf of [insert the name of the
architect/practice in the contract]


Copies: Employer
        Quantity surveyor
        Consultants
        Clerk of works
```

Fig 4.6

Letter from architect to employer: if disagreeing with former architect's decisions, etc.

```
PROJECT TITLE

Dear Sir

I have now had the opportunity of examining all the
drawings, files and other papers relating to this
contract and I have visited the site and spoken to
the contractor.

Article 3 of the contract prevents me from
disregarding or overruling any certificate or
instruction given by the architect previously engaged
to administer this contract.

Purely for record purposes, therefore, I list below
the matters on which I find myself unable to agree
entirely with the decisions of the former architect:

[list all matters with which you disagree]

When you have had the opportunity to study these
matters I suggest we should meet to discuss ways of
dealing with them.

Yours faithfully
```

If the named architect dies or ceases to act for some other reason, it is the employer's duty to nominate a successor within 14 days of the death or ceasing to act of the named architect: *Croudace Ltd* v *London Borough of Lambeth* (1986) 6 ConLR 72. Unlike the position under JCT 80, there is no provision for the contractor to object to the successor, but there is nothing to prevent the contractor from referring the matter to immediate arbitration if he feels strongly about it. The contractor must, however, continue with the works while awaiting the outcome of the arbitration and, in the majority of contracts carried out under this form, it is likely that the work will be complete by the time a decision is reached. That is probably the reason for the exclusion of the right of objection in the first place. Clearly, the employer would do well to listen carefully if the contractor makes any representations, in order to avoid difficulties.

There is an important proviso to this Article which states that 'no person subsequently appointed to be the Architect . . . under this Contract shall be entitled to disregard or overrule any certificate or instruction given by' the former architect. This is a sensible provision to safeguard the contractor's interests under circumstances which are in the sole control of the employer. If you are the successor and you disagree with some of your predecessor's decisions, you should notify the employer immediately, (Fig 4.6), to safeguard your own position, even though you cannot alter them.

Although clause 1.1 is titled *Contractor's obligation*, it contains a trap for the unwary architect. After concisely stating the contractor's obligation to carry out and complete the works with due diligence and in a good and workmanlike manner in accordance with the contract documents, there is a proviso:

'where and to the extent that approval of the quality of materials or of the standards of workmanship is a matter for the opinion of the Architect . . . such quality and standards shall be to the reasonable satisfaction of the Architect . . .'

What the proviso means, in essence, is that, if you reserve anything to do with quality and standards as a matter for your opinion, they must be to your reasonable satisfaction. You have considerable scope, but you are to be reasonable about it. If the contractor disagrees that you are being reasonable, he can, of course, seek arbitration on the point. The danger is threefold:

○ The proviso implies that if you do reserve matters to your approval, you will, in due course and probably when the contractor requests your decision, give your approval. The contract provides no machinery for so doing, but the wise contractor will ask you to give approval in writing and it is difficult to see how you could be justified in refusing unless you do not approve.

○ The final certificate is not stated to be conclusive as regards anything, thus avoiding the unquestionable trap contained in JCT 80, but if you have reserved any matter to your approval, it is reasonable to assume that you would express your disapproval (if appropriate) before you make yourself *functus officio* by the issue of the final certificate.

○ If you formally approve a matter you have so reserved, your approval will override any requirement in the contract documents, so if you approve something which is not in accordance with the contract, the employer is prevented from seeking redress from the contractor on the ground of lack of compliance.

It is clearly wise to limit the matters to which you wish to reserve your approval. The use of some such common phrase as 'unless otherwise stated, all workmanship and materials are to be to the approval of the architect' should be strictly avoided because it will put an enormous burden of approval on you throughout and/or at the end of the contract. Clause 1.2 very briefly sets out your duties. These are:

○ The issue of further information.

○ The issue of all certificates in accordance with the contract (see Table 4.7)

○ The confirmation of all instructions in writing in accordance with the contract.

Table 4.7
Certificates to be issued by the architect under MW 80

Clause	Certificate
2.4	Practical Completion
2.5	Making Good of Defects
4.2	Progress Payments
4.3	Penultimate Certificate
4.4	Final Certificate

The amount, if any, of further information necessary will vary greatly depending upon the size and complexity of the work. In the case of small simple projects, the contract drawings may need very little elaboration. Whether the information is necessary should be a matter of fact, not opinion. Note that there is no obligation on the contractor to apply for further information either in this clause or anywhere else in the contract. The prudent contractor will do so, but it is your duty to supply the information to enable him to carry out the works properly in accordance with the contract. The completion date is part of the provisions of the contract, therefore your duty is to supply necessary information at the correct times to enable the contractor to proceed with the works and to complete them by the contract completion date. If you fail to do so, it will be a breach for which the contractor can claim extension of time and possibly damages at common law.

The issue of certificates refers not only to progress payments but also to such things as practical completion and making good of defects after the defects liability period. Although there is no prescribed way in which a certificate should be set out, it is a formal document. It may be in the form of a letter, but for the avoidance of doubt it is always good practice to head the letter 'Certificate of . . .' and begin 'I certify . . .' Where a number of certificates are to be issued in sequence it is normal to number them in order. Delay in the issue of a certificate is a serious matter and may give rise to financial claims by the contractor (Table 10.4).

Confirming instructions will be dealt with in 4.3 below.

The remaining express duties of the architect are covered in the appropriate chapters of this book. Because this contract is so brief, your duties are briefly stated therein. This can be misleading and you should always bear in mind that the courts will imply terms to cover the way in which you must administer the contract. Generally, you are expected to act promptly and efficiently. If you do, you will avoid claims and contribute towards the smooth running of the contract.

4.3   Architect's instructions

All your instructions to the contractor must be confirmed in writing to be effective. Despite what some contractors maintain, there is absolutely no requirement that instructions must be on a specially printed form headed 'Architect's Instruction' although it is undoubtedly good practice to use such forms because:

○   It leaves no room for doubt that you are issuing an instruction. An instruction included in a letter dealing with a great many other things can sometimes be confusing.

○   It makes the job of keeping track of instructions for the purposes of valuation and checking so much easier.

Instructions can be given in letter form, provided you make it clear that you are issuing an instruction (Fig 4.8). You can also give a hand-written instruction on site, provided it is signed and dated (always sign on behalf of the architect named in the Articles if it is someone other than yourself). Instructions contained in the minutes of site meetings are valid if you are the author of the minutes and if they are recorded as agreed at a subsequent meeting. It is probable, however, that such instructions are not effective until the contractor receives a copy of the minutes recording agreement. Since site meetings are sometimes at monthly intervals, the matter giving rise to the instruction may be ancient history before the contractor's duty to comply becomes operative; minuted instructions are, therefore, best avoided.

The position with regard to drawings is uncertain. If you issue a drawing together with a letter instructing the contractor to use the drawing for the works, that is certainly an instruction for contract purposes. If you simply issue a drawing under cover of a compliments slip, the drawing may be an instruction or it may simply be sent for comment. The contractor would be wise to check with you first, but you may have difficulty showing that the drawing was not intended as an instruction if the contractor simply carries out the work. Note that if you send the contractor a copy of the employer's letter requiring some action under cover of a compliments slip, it may not be an instruction at all. The moral is clear – never use compliments slips.

Clause 3.5 is the principal contract provision governing the issue of instructions. The procedure is shown in Flowchart 4.9. Despite what, at first sight, may appear to be an all-embracing provision, 'The Architect . . . may issue written instructions which the Contractor shall forthwith carry out', you must act within the scope of your authority and the instruction must relate to the contract works. The courts do not generally favour sweeping generalisations and prefer to see respective rights and duties expressed in precise terms. In the event of a dispute, therefore, it is probable that the clause will be given a very narrow interpretation. The contractor must carry out the instruction as soon as reasonably can be (*London Borough of Hillingdon* v *Cutler* [1967] 2 All ER 361).

Provision is made for the situation if instructions are issued orally. They must be confirmed in writing by the architect within two days. The contract is silent as to when such an instruction becomes effective. It is considered that, in context, the instruction would not become effective until the contractor receives the confirmation. Although the practice of issuing oral instructions is widespread, it is difficult to see why it was necessary to make special provision in the contract since the effect of the confirmation is not different from the effect of a simple written instruction in the first place. The oral instruction itself is irrelevant. Note that there is no provision for the contractor to confirm oral instructions. It is considered that the contractor's attempts to confirm will be ineffective, even if you do not dissent, provided that you do not acknowledge receipt. Ideally, oral instructions should be avoided.

If the contractor does not carry out your instructions 'forthwith' the contract provides a remedy in much the same form as JCT 80, clause 4.1.2. As a first step, you must send the contractor a written notice requiring him to comply with the instruction within 7 days (Fig 4.10). If the contractor does not so comply, you should advise the employer (Fig 4.11) that he may employ others to carry out the work detailed in the instruction. Although the contract specifically states that it is the employer who may employ others, he will expect you to advise him and handle the details. It will amount to a completely separate contract and, in order to ensure that

Fig 4.8
Letter from architect to contractor: issuing an instruction

```
PROJECT TITLE

Dear Sir

In accordance with clause 2.5/3.4/3.5/3.6/3.7/6.3B
[delete as appropriate] please carry out the
following instruction forthwith:

[insert the instruction]

Yours faithfully
```

Flowchart 4.9
Architect's instructions

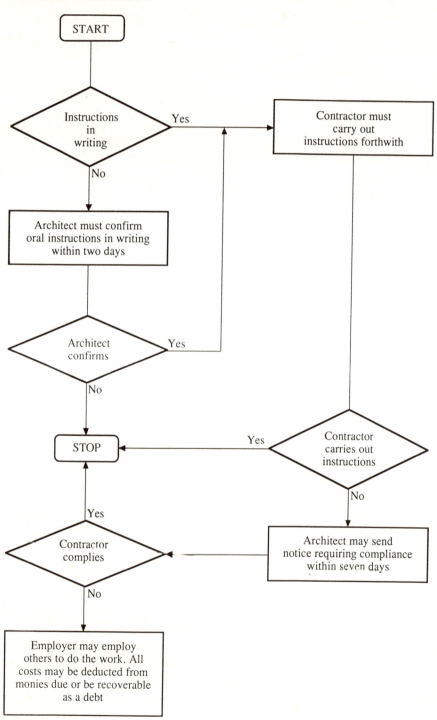

Fig 4.10

Letter from architect to contractor: requiring compliance
with instructions before default action taken

RECORDED DELIVERY *

                                                  Date

PROJECT TITLE

Dear Sir

On [insert date] I instructed you to [specify] under
clause 3.5 of the contract. That clause requires
you to carry out my instructions forthwith.

You have failed to so comply and in accordance with
clause 3.5 I hereby require you to carry out the
above-mentioned instruction. Should you fail to
comply within 7 days after receipt of this notice,
the employer will engage others to carry out the work
and deduct all the resultant costs from monies due to
you under the contract.

Yours faithfully

Copies: Employer
        Quantity surveyor (if appointed)

*Service by recorded delivery is not required by
 clause 3.5 but is a desirable precaution.

Fig 4.11

Letter from architect to employer: if contractor fails to comply with instruction within 7 days after notice

PROJECT TITLE

Dear Sir

I refer to my letter to [*insert name of contractor*], dated [*insert date*], requiring him to comply with my instruction of the [*insert date*] within 7 days.

The 7 days expired yesterday and, during a site visit this morning, I observed that the contractor had not complied with my instruction.

I advise that you should take advantage of the remedy afforded by clause 3.5 of the Conditions of Contract and, if you will let me have your written instructions to that effect, I will obtain competitive tenders from other firms for the carrying out of the work. All additional costs, including my additional fees, can be deducted from future payments due to the contractor and I will liaise with you regarding these details at the appropriate time.

Yours faithfully

there can be no reasonable question about costs, you should obtain competitive quotations from three firms if time and circumstances permit. After the work is completed, the employer has the right to deduct all costs from monies due or to become due to the contractor under the contract. Alternatively, but less attractively, the monies may be recovered as a debt. In some instances this may be the only practicable way of obtaining the cash, but it is best avoided if possible because it involves the always uncertain and costly process of litigation. Note that the 'costs incurred thereby' refer to additional costs, over and above the cost of the instruction. It is not intended that the employer should get the work done at the contractor's expense. The additional costs will include such items as scaffolding, cutting out and making good, depending upon circumstances. You are entitled to charge additional fees and the employer may include them in his computation of costs together with any other incidental expenses attributable to the contractor's non-compliance. When deducting the amount, the contractor is entitled to a brief statement showing how the figure has been calculated. Although the wording in JCT 80 is slightly different, it is not thought that it makes any practical difference. Note that there is no provision for the contractor to object to any instruction or to request the architect to state the empowering provision. Apart from clause 3.5, there are only five other clauses empowering the architect to issue instructions in particular circumstances:

o   Clause 2.5 empowers the architect to instruct the contractor not to make good defects at his own cost (this provision is discussed in detail in Chapter 9).

o   Clause 3.4 empowers the architect to issue instructions requiring the exclusion from the works of any person employed thereon. The instruction is not to be issued unreasonably or vexatiously and is clearly intended to enable the architect to have incompetent operatives removed. Although the point is debatable, we incline to the view that the clause, unlike the similar provision in JCT 80, does not permit the exclusion of a firm because that is not the ordinary meaning of 'person' (which in JCT 80 is extended by clause 1.3).

o   Clause 3.6 empowers the ordering of variations by way of additions, omissions or other changes in the works or the order or period in which they are to be carried out.

o   Clause 3.7 requires the architect to issue instructions regarding the expenditure of any provisional sum (see Chapter 11).

o   Clause 6.3B requires the architect to issue instructions for the reinstatement and making good of loss or damage caused by fire, lightning, storm, etc. (see Chapter 3).

Clause 3.3 provides that any instructions given to the contractor's representative on the works are deemed to have been issued to the contractor. In the context of the contract as a whole, such instructions must be in writing for the clause to have any effect although, in practice, the architect seldom gives written instructions to the person in charge. Site instructions tend to be oral, therefore ineffective, written instructions being sent to the contractor's main office.

It has already been mentioned that the courts will probably construe clause 3.5 narrowly as regards your power to issue instructions. Although not expressly mentioned, it is considered that the following common situations will fall within your powers under the clause:

o   The correction of inconsistencies between the contract documents. Although clause 4.1 deals with this matter, it does not specifically provide that you can issue instructions. None the less, it is considered that you certainly have that power.

o   Opening up of work for inspection and testing of materials. If the work or materials are found to be in accordance with the contract, the contractor will have a claim for extension of time. It is thought that, under those circumstances, the costs of carrying out the instruction could be valued under clause 3.6, but if the contractor considers that he has been involved in loss and/or expense, that would be the subject of a common law claim (see Chapter 10).

o   The removal or correction of defective work. The clause appears to be broad enough to cover either requirement.

o   The postponement of any work in progress. This power does not extend to deferring the giving of possession of the site which would amount to varying an express term of the contract – a matter for the parties to negotiate. Any postponement instruction would inevitably give rise to an extension of time, a claim for damages at common law and, if the suspension extends to the whole of the works and lasts for a continuous period of at least a month, would provide a ground for determination by the contractor under clause 7.2.3.

4.4   Summary

Authority

o   From the contractor's point of view, your authority is defined by the terms of the contract.

o   The contractor can only take action against you in tort.

o   The employer can take action against you in contract and in tort.

o   Your authority depends on what you have agreed with the employer.

o   You are expected to use the same degree of skill as any other average, competent architect unless you profess a higher degree of skill in a particular area.

o   Once the contract is signed, you have a responsibility to both parties.

o   You do not have a quasi-arbitral role nor are you immune from actions for negligence as a result of your decisions.

o   If you are the employee of the employer, you may be in a difficult position.

o   You cannot overrule the decision of a previous architect engaged on the work.

o   Beware of leaving matters to your approval.

o   Your approval will generally override contract requirements.

o   All instructions must be confirmed in writing by you.

o   You have implied as well as express duties.

o   Your general power to issue instructions is confined to matters relating to the works.

Chapter 5

Contractor

5.1 Contractor's obligations: express and implied

5.1.1 Legal principles

It is all too commonly assumed that the whole of the contractor's contractual obligations are contained and set out in the printed contract form. This is not so and the contract form is silent on many important matters. These gaps are filled in by terms which the law will write into the contract. If nothing at all were said about the contractor's obligations, the law would require him to do three things:

○   To carry out his work in a good and workmanlike manner exercising reasonable care and skill. This means that the contractor must show the same degree of competence as the average contractor experienced in carrying out that type of work.

○   To use materials of good quality which are reasonably fit for their purpose.

○   To ensure that the completed building or structure is reasonably fit for its intended purpose. This obligation is modified where the employer engages an architect to design the works since the architect is then responsible for the design.

Similarly, a term would be written in that the contractor would comply with the building regulations and other statutory requirements – a matter which is the subject of an express term in MW 80

These implied terms can be modified by the express terms of the contract itself and are, in fact, modified by what MW 80 says, with the result that the contractor is under a lesser duty than would otherwise be the case.

However, statute also imposes similar implied obligations on the contractor and in particular the Supply of Goods and Services Act 1982 implies terms as to the quality of goods supplied by the contractor under the contract. The construction of dwellings (both houses and flats) is governed by the Defective Premises Acts 1972 which provides:

'Any person taking on work for or in connection with the provision of a dwelling . . . owes a duty to see that the work that he takes on is done in a workmanlike manner, with proper materials . . . and so as to be fit for the purpose required . . .'

The express provisions of MW 80 must, therefore, be read against this background. There is a further important point. As discussed in Chapter 8, clause 3.2 envisages that the contractor may sub-let the works or any part of them if the architect gives his written consent. Sub-contracting does not free the contractor from responsibility for the sub-contracted work. The contractor remains responsible to the employer for all defaults of the sub-contractor as regards workmanship, materials or otherwise. Tables 5.1 and 5.2 summarise the contractor's powers and duties respectively under the express provisions of MW 80.

5.1.2 Execution of the works

Clause 1.1 requires the contractor 'with due diligence and in good and workmanlike manner (to) carry out and complete the Works in accordance with the Contract Documents . . ' which are specified in the first recital. This is his basic and absolute obligation and seems to equate with the common law duty. The contractor is bound to complete the Works by the date for completion set out in clause 2.1 come what may. He must bring the Works to a state when they are 'practically completed' so that you can issue your certificate under clause 2.4. Although the contractor's basic obligation is not qualified in any way in clause 1.1, it is subject to two qualifications:

o   The provisions of clause 7.2 entitling the contractor to determine his employment under the contract, e.g., if the employer suspends the carrying out of the Works for a continuous period of at least one month, and subject to the procedural requirements of that clause.

o   If the employer – or you on his behalf – prevents the contractor from completing his work by the completion date unless, of course, a proper extension of time has been validly made.

The contractor must carry out his work *in accordance with* the contract documents as defined. You must take care to ensure that the description of the work is adequate and this means using precise language. All too often important parts of the contract documents are written in slovenly English. It is impossible to enforce generalisations. The contract documents must contain all the requirements which the employer wishes to impose, and you should avoid the use of phrases such as 'of good quality' or 'of durable standard'. Remember, too, that you cannot use the contract documents to

Table 5.1
Contractor's powers under MW 80

Clause	Power	Comment
2.1	Commence the works on the specified date	
3.1	Assign the contract	If the employer gives written consent
3.2	Sub-let the works or part thereof	Only if the architect consents in writing. Consent must not be unreasonably withheld
3.6	Agree the price of variations	○ If the architect consents **and** ○ Before executing the variation instruction
7.2	Forthwith determine the employment of the contractor by notice	The notice must not be given unreasonably or vexatiously and must be served on the employer by registered post or recorded delivery. Notice can be served: ○ If the employer fails to make any progress payment due within 14 days of payment being due **and** ○ If the default has continued for 7 days after receipt of a notice from the contractor specifying the default, served by registered post or recorded delivery post OR ○ If the employer or any person for whom he is responsible interferes with or obstructs the carrying out of the works **and** ○ The default has continued likewise OR ○ If the employer fails to make the premises available to the contractor on the specified date **and** ○ If the default has continued likewise OR ○ If the employer suspends the carrying out of the works for a continuous period of at least one month **and** ○ If the default has continued likewise OR ○ If the employer becomes insolvent

Table 5.2
Contractor's duties under MW 80

Clause	Duty	Comment
1.1	Carry out and complete the works in accordance with the contract documents in a good and workmanlike manner and with due diligence	
2.1	Complete the works by the specified date	The architect must insert the date
2.2	Notify the architect of delay	If: ○ It becomes apparent that the works will not be completed by the specified date **and** ○ This is because of reasons beyond the control of the contractor
2.3	Pay liquidated damages to the employer	If the works are not completed by the specified date or by any later date fixed under clause 2.2
2.5	Make good at his own cost any defects, excessive shrinkages or other faults	The defects, shrinkages or other faults must have appeared within 3 months of the date of practical completion **and** must be due to materials or workmanship not in accordance with the contract or to frost occurring before practical completion **and** unless the architect has instructed otherwise
3.3	Keep a competent person in charge upon the works at all reasonable times	Under clause 3.4 the architect may instruct this person's exclusion
3.5	Carry out all architect's instructions forthwith	The instructions must be in writing
4.4	Supply the architect with all documentation reasonably necessary to enable the final sum to be computed	○ Within three months of the date of practical completion ○ It is not conditional upon a request from the architect
5.1	Comply with and give all notices required by any statute, etc.	
5.1	Pay all fees and charges in respect of the works	Provided they are legally recoverable from him

Clause	Duty	Comment
5.1	Give immediate written notice to the architect specifying any divergence	If the contractor finds any divergence between the statutory requirements and the contract documents **or** between such requirements and an architect's instruction
5.4	Comply with the fair wages resolution	If applicable – local authority contracts
6.1	Indemnify the employer against any expense, liability, loss, claim or proceedings in respect of personal injury or death	If the expense, etc., arises out of or in the course of or is caused by reason of the carrying out of the works unless due to any act or neglect of the employer or those for whom he is legally responsible
6.1	Maintain and cause his sub-contractors to maintain the insurances necessary to meet his liability under clause 6.1	
6.2	Idemnify the employer against and insure and cause his sub-contractors to insure against any expense, liability, loss, claim or proceedings for damage to property	This is subject to clauses 6.3A or 6.3B. The indemnity operates if the expense, etc.: ○ Arises out of or in the course of or is caused by reason of the carrying out of the works **and** ○ Is due to any negligence, omission or default of the contractor or any sub-contractor or any person for whom they are legally responsible
6.3A	Insure against the specified risks	New works only
6.3A	Restore or replace work or materials, etc., dispose of debris, and proceed with and complete the works	Upon acceptance of a clause 6.3A claim by the insurers
6.4	Produce evidence of insurances and cause sub-contractors so to do	If the employer so requires
7.1	Immediately give up possession of the site	Where the employer determines the contractor's employment

override or modify the printed conditions because of the last sentence of clause 4.1.

The contractor must complete *all* the work shown in, described by or referred to in the contract documents. His obligation only comes to an end when you have issued your certificate of practical completion. Thereafter, the contractor must fulfil his obligations under the defects liability clause (see section 9.3.1).

The proviso to clause 1.1 states that if approval of workmanship or materials is a matter for your opinion, then quality and standards must be to your reasonable satisfaction. The effect of this is discussed in Chapter 4, section 4.2.

There is no further amplification of the contractor's basic obligation although he is, of course, bound to complete the Works *by* the date stated in clause 2.1 and if he does not do so, subject to the operation of the provisions for extension of time, he must pay liquidated damages to the employer.

But by implication he is under a further obligation in progressing the Works. He must proceed with the works 'regularly and diligently' because his failure to do so *without reasonable cause* is one of the grounds which may entitle the employer to bring his employment under the contract to an end under clause 7.1.1.

Unfortunately, there is no clear-cut decision on what is meant by proceeding 'regularly and diligently'. 'Going-slow' is certainly not a breach of contract as such (*J. & M. Hill & Son Ltd* v *London Borough of Camden* (1980) 18 BLR 31) nor are temporary defects in the work. At the end of the day, whether or not the contractor is proceeding regularly and diligently is a question of fact to be judged by the standards of the industry using the average competent and experienced contractor as a yardstick. It has, in any event, been suggested that the obligation to proceed regularly and diligently is a minor obligation and subsidiary to the principal obligation. Sensibly you, as architect, should require the contractor to provide you with a programme or progress schedule as a yardstick against which the contractor's progress can be measured. A programme or progress schedule is not a contract document but is nevertheless a valuable aid. Conversely, you should beware of approving any programme which the contractor may submit to you in case he tries to rely upon your approval if things do not go as planned.

5.1.3 Workmanship and materials

Clause 1.1 also deals with workmanship and materials. It requires the contractor to use materials and workmanship 'of the quality and standards . . . specified' by you in the contract documents. This in fact imposes an obligation on you to define the quality and standards of workmanship and materials very carefully indeed and where it is impossible adequately to specify in sufficient detail, you could do worse than fall back on the phraseology of the common law:

○   All work shall be carried out in a good and workmanlike manner and with proper care and skill.

○   All goods and materials shall be of good quality and reasonably fit for their intended purpose.

Quite clearly, the contractor is expected to show a reasonable degree of competence and to employ skilled tradesmen. Although you have no power to direct him how he should carry out his work, save to the extent that you can by a variation order issued under clause 3.6 change the order or period in which the works are to be carried out, you do have power under clause 3.4 to instruct that any unsatisfactory employee or anyone else employed on the Works should be excluded.

Because of the wording of clause 1.1 the contractor *must* provide workmanship, materials and goods of the standards and quality specified. It is no excuse that they are not in fact available, since he does not enjoy the benefit of the limitation of JCT 80 that he need do so only so far as the goods, etc., are procurable. This is a matter which is at the contractor's risk.

Materials and goods are also referred to in clause 4.2 in the context of progress payments. You are *bound* to include in progress certificates 'the value of any materials or goods which have been reasonably and properly brought upon the site for the purpose of the works and which are adequately stored and protected against the weather and other casualties'. This is a very dangerous provision from the employer's point of view since the unfixed materials and goods will not necessarily become the property of the employer, even though he has paid for them, if, in fact, they are not his property in law, as will often be the case. Builders' merchants frequently include a 'retention of title' clause in their contracts of sale: *Dawber Williamson Roofing Co Ltd* v *Humberside County Council* (1979) 14 BLR 70.

Ideally, clause 4.1 should be amended so as to ensure that the inclusion of the value of unfixed materials is a matter for your discretion. Less satisfactorily, it may include an appropriate provision requiring the contractor's applications for progress payments to be accompanied by documentary proof of ownership.

It is probable that you are safe from an action for negligence if you do operate clause 4.1 as it stands because the employer has signed MW 80 which says that you *shall* include on-site unfixed materials, but you should be alert to the dangers, and the Joint Contracts Tribunal should issue an amended clause 4.2 as a matter of urgency.

5.1.4 Statutory obligations

MW 80 clause 5.1 imposes on the contractor a duty to comply with all statutory obligations, e.g., those imposed by the Building Regulations 1985 and to pay all fees and charges which are legally recoverable from him. It also imposes on him an obligation to give you immediate written notice if he discovers any divergence between the statutory requirements and the contract documents or one of your instructions.

There then follows a most curious provision:

'Subject to this latter obligation, the contractor shall not be liable to the employer under this contract if the Works do not comply with the statutory requirements where and to the extent that such non-compliance . . . results from the contractor having carried out work in accordance with the contract documents or any instruction of the architect'.

The effect of this provision is contractually to exempt the contractor from liability to the employer if the works do not comply with statutory requirements provided he has carried out the work in accordance with the contract documents or any instruction of yours, where, for example, he does not spot the divergence. His obligation to notify you of a divergence only arises *if* he spots it, and unless he does so he is under no obligation to notify you. Plainly, it is ineffective to exonerate him from his duty to comply with statutory requirements and if, for example, he carried out work in breach of the Building Regulations, he would be criminally liable since the primary liability to comply with them rests on him: *Perry v Tendring District Council* (1985) 3 ConLR 74. Although the contrary view has been expressed about the similarly-worded provision in JCT 80, the wording is probably sufficiently wide to protect the contractor from any action by the employer. But it does not protect you, and if the fault is yours the employer will be able to recover his losses from you.

Despite the fact that the Fair Wages Resolution has been rescinded by the House of Commons, where the local authority is the employer, clause 5.4 obliges the contractor to comply with its conditions, now set out in Part D of the Supplementary memorandum, because the standing orders of some local authorities still require compliance. In current economic conditions clause 5.4 is a dead letter.

5.1.5 Contractor's representative

Clause 3.3 obliges the contractor to keep on the works a 'competent person in charge' and he must do this at all reasonable times, i.e., during normal working hours. This person is intended to be the contractor's full-time representative on site, but his appointment and replacement are not subject to your approval although in an appropriate case you could exercise your powers under clause 3.4 to require his exclusion from the site and thus force the contractor's hand.

Competent means that the contractor's representative must have sufficient skill and knowledge, and it is essential that you be aware from the outset of this person's identity since he is the contractor's agent for the purpose of accepting your instructions. Instructions given to the person in charge are *deemed* to be given to the contractor.

Many architects include in the tender documents a requirement that the contractor give adequate notice of the replacement of the person-in-charge.

5.2   Other obligations

5.2.1   Access to the works and premises

There is no express provision in MW 80 for you and your representatives to have access to the works at all reasonable times. Such a right of access to the site is implied under the general law. However, as is the case under IFC 84, there is a gap, because you may need access to the contractor's workshops, etc., where items are being prepared for the contract. If you do need to visit the contractor's premises to keep an eye on things, you should include an appropriate clause in the contract documents, following the wording of JCT 80 clause 9, because it is probable that you do not have that right of access under the general law.

5.2.2   Compliance with architect's instructions

Clause 3.5 obliges the contractor to carry out your written instructions forthwith and this obligation is not conditioned in any way. The contractor is given no right to object to your instructions, even those involving a variation.

The sanction for non-compliance by the contractor is set out in the second paragraph of clause 3.5. If a contractor fails to comply with one of your instructions, you should serve the contractor with a written notice requiring compliance. If the contractor has not complied within 7 days from receipt of that notice, the employer may engage others to carry out the work and deduct the cost from monies due to the contractor. Fig 4.10 is a suggested letter.

To avoid the possibility of an argument that the employer has waived his rights, it is essential that you ensure that the machinery provided is put into operation if the contractor fails to comply with instructions.

You should note that MW 80 contains no express provision for you to issue instructions for the removal of work not in accordance with the contract as does JCT 80 clause 8.4, but it is probable that the broad wording of clause 3.5 extends to cover that situation.

5.2.3 Other rights and obligations

Tables 5.1 and 5.2 summarise the contractor's rights and duties generally. Other matters referred to in those Tables are dealt with in the appropriate chapters.

5.3   Summary

Contractor's obligations

MW 80 imposes certain express obligations on the contractor; other obligations are implied at common law.

o   The contractor must carry out and complete the works with due diligence and in accordance with the contract documents.

o   He must do this by the specified completion date.

o   The contract documents must be precise and specify quality and standards of both workmanship and materials.

o   The contractor must comply with statutory obligations and pay all fees and charges involved.

o   *If* he discovers a divergence between the statutory requirements and the contract documents or one of your instructions he must give you written notice immediately.

o   Subject to this he is exempted from liability to the employer if the works as built contravene statute law.

o   The contractor must keep a competent representative on site during normal working hours and any instructions of yours given to that representative are deemed to have been given to the contractor.

o   The contractor must comply with all your instructions issued under the contract.

o   The employer has an option to engage others to carry out your instructions should the contractor not carry them out within 7 days of a written notice from you.

Chapter 6

Employer

6.1   Powers and duties: express and implied

Like those of the contractor, some of the employer's powers and duties arise from the express provisions of MW 80, while others arise under the general law.

Tables 6.1 and 6.2 set out those powers and duties of the employer which arise from the express terms of the contract.

This section is concerned with the obligations which are placed on the employer by way of implied terms. These are provisions which the law writes into a contract in order to make it commercially effective and terms will be implied to the extent that they are not inconsistent with the express terms which may also exclude or modify the implied terms. MW 80 does not, in fact, modify or exclude two important implied terms.

It is an implied term of MW 80 that the employer will do all that is reasonably necessary on his part to bring about completion of the contract. Conversely, it is implied that the employer will not so act as to prevent the contractor from completing in the time and in the manner envisaged by the agreement. Breach of either of these implied terms which results in loss to the contractor will give rise to a claim for damages at common law and the contractor can pursue that claim under the arbitration agreement.

Equally, if the employer – either personally or through your agency or that of anyone else for whom he is responsible in law – hinders or prevents the contractor from completing in due time, he is not only in breach of contract, but will be disentitled from enforcing the liquidated damages clause.

The various cases put the duty in different ways, but in essence the position is summarised:

○   The employer and his agents must do all things necessary to enable the contractor to carry out and complete the works expeditiously and in

Table 6.1
Employer's powers under MW 80

Clause	Power	Comment
3.1	Assign the contract	If the contractor consents
3.5	Employ and pay others to carry out the works	If the contractor fails to comply with a written notice from the architect requiring compliance with an instruction and within 7 days from its receipt
5.5	Cancel the contract and recover resulting loss from the contractor	If the contractor is guilty of corrupt practices as specified in the clause
7.1	Determine the employment of the contractor by written notice	If the contractor ○ Fails to proceed diligently with the works without reasonable cause OR ○ Wholly suspends the carrying out of the works before completion OR ○ Becomes insolvent, etc. The determination notice must not be served unreasonably or vexatiously

accordance with the contract.

○   Neither the employer nor his agents will in any way hinder or prevent the contractor from carrying out and completing the works expeditiously and in accordance with the contract.

The scope of these implied terms is very broad and in recent years more and more claims for breach of them have been before arbitrators and the courts. The employer must not, for example, attempt to give direct orders to the contractor, and he must see that the site is available to the contractor on the date specified in clause 2.1 and that access to it is unimpeded by those for whom he is responsible. This is especially important in the case of work to existing structures or occupied buildings, and these are matters which you should discuss with the employer before the contract is let.

6.2   Rights under MW 80

6.2.1 General

Although the contract is between the employer and the contractor – who are the only parties to it – an analysis of the contract clauses shows that the employer has few rights of any substance.

Table 6.2
Employer's duties under MW 80

Clause	Duty	Comment
4.2	Pay to the contractor amounts certified by the architect	Payment must be made within 14 days of the date of the certificate issued under clause 4.2
4.3	Pay similarly	If the architect issues a certificate under clause 4.3
5.2	Pay to the contractor any VAT properly chargeable	
6.3A	Pay insurance monies to the contractor	On certification under clause 4.0
6.3B	Maintain adequate insurances against the specified risks	Existing structures only
6.4	Produce evidence of such insurances	If the contractor so requires
7.2	Pay to the contractor such sum as is fair and reasonable for the value of work begun and executed, materials on site, and removal costs as specified	If the contractor determines his employment under clause 7.2

His major right is to have the work contracted for handed over to him by the agreed completion date, properly completed in accordance with the contract documents. He also has the right to assign the benefits of the contract to a third party, provided the contractor consents in writing (clause 3.1) although it is difficult to envisage many circumstances in which an employer could wish to do that!

6.2.2 Damages for non-completion

If the contractor does not complete the works by the agreed completion date, the employer is entitled to recover liquidated damages at the rate specified in clause 2.3 for each week or part of a week during which the works remain uncompleted after the original or extended completion date.

There is no requirement that you issue a certificate of non-completion or any other precondition. The mere fact of late completion is sufficient to bring clause 2.3 into operation. Although the employer is given no express right to deduct liquidated damages from monies due to the contractor, a right of deduction is implied, but if (unusually) no sums are due or to become due to the contractor, the employer must sue for them as a debt.

You should, of course, advise the employer of his rights should the contract overrun.

6.2.3 Other rights

These are summarised in Table 6.1 and are there described as 'powers'. They are discussed in the appropriate chapters.

6.3   Duties under MW 80

6.3.1 General

The essence of a duty is that it must be carried out. It is not permissive; it is mandatory. Breach of a duty imposed by the contract will render the employer liable to an action for damages by the contractor in respect of any proven loss.

Some breaches of contract may, in fact, entitle the contractor to treat the contract as at an end. Lawyers call them 'repudiatory breaches' which means that they go to the basis of the contract. For example, physically expelling the contractor and his men from site would be a repudiatory breach since the employer is effectively showing that he no longer wishes to be bound by the contract.

Breach of any contractual duty will always, in theory, entitle the contractor to at least nominal damages, although in many cases any loss will be difficult if not impossible to quantify.

6.3.2 Payment

From the contractor's point of view, the most fundamental duty of the employer is to make payment in accordance with the terms of the contract. However, while steady payment against certificates is essential from the contractor's view-point, the general law does not regard none or late payment as such as a major breach of contract, and recent case law suggests that a certificate is not as good as cash, as most contractors appear to think, since in some limited cases the employer may be justified in withholding payment of certificated amounts pending arbitration: *C M Pillings & Co Ltd* v *Kent Investments Ltd* (1986) 4 ConLR 1.

MW 80 is quite specific in its terms as to payment (see Chapter 11). The basic provisions are to be found in clauses 4.2, 4.3 and 4.4. The contractor's remedies for non-payment are limited. He can sue for payment on the certificate once payment is due and may seek summary judgment. However, he is not entitled to interest on the overdue sum as damages

(*President of India* v *La Pintada Compania Navigaçion SA* [1984] 2 All ER 773) although the courts have power to award interest from the date when payment should have been made until the date of judgment. The contractor certainly has no right to suspend work because he is not being paid, but he does have the right to determine his employment under the contract under clause 7.2.1 in respect of the employer's failure to make progress payments.

If the employer is a tardy payer, you should make him aware of the need to pay promptly in accordance with the contract terms because nothing sours good working relationships more than late payment, and most contractors do have a cash-flow problem (Letter 6.3).

Once you have issued a payment certificate under clauses 4.2 and 4.3 (progress payments and penultimate certificate) the employer is given a period of grace in which to honour the certificate. He must make payment *within 14 days of the date of the certificate*, which means payment before the expiry of that period. This is a very tight timetable and you must send the certificate to the employer on the day you issue it. Assuming no postal delays (which cannot be relied on these days) and dispatch by first-class post, the employer effectively has no more than 13 days for paying what is due. The same timetable effectively applies to the final certificate under clause 4.4, though the wording in that case is that the sum due is a debt 'as from the fourteenth day after the date of the final certificate'.

Payment by cheque is probably good payment, but it is no excuse for the employer to say, for example, that his computer arrangements do not fit in with the scheme of certificates, which is an increasingly common excuse for late payment. If this is indeed the case the payment period should have been amended before the contract was let.

The employer is entitled to retain retention on progress payments and, in the case of a contract overrun, can set-off amounts due to him as liquidated damages under clause 2.3.

6.3.3 Retention

The employer has certain rights in the retention percentage, which is commonly 5%. It is not stated that the retention is trust money and it may be, therefore, that the contractor is at risk if the employer becomes insolvent.

The employer's rights in the retention are to have it as a fund from which he is entitled to have defects remedied and other bona fide claims settled. But morally the retention is the contractor's money and it would be an improvement if the JCT would amend clause 4.2 to provide that the employer's interest in the retention was 'fiduciary as trustee for the contractor'.

Fig 6.3

Letter from architect to employer: if employer is slow to honour certificates

```
PROJECT TITLE

Dear Sir

The contractor has complained to me that he is not
receiving payment due on certificates until after the
period allowed in the contract.

Under the terms of the contract, you have fourteen
days from the date of the certificate within which to
make payment.  Failure to pay within the stipulated
time entitles the contractor to determine his
employment under the contract or to seek summary
judgment.  If the contractor determined his
employment, the consequences would be extremely
expensive.

Disregarding the strict legal requirements as to
payment, it is good practice to pay promptly because
the contractor is always in the position of having
paid out substantial sums well before payment is due.
Prompt payment is crucial to his cash flow and,
consequently, late payment spoils good working
relations.

If you would bear this in mind, there should be
benefits on all sides.

Yours faithfully
```

6.3.4 Other duties

These are summarised in Table 6.2 and are commented on as appropriate in other chapters.

6.4   Summary

Rights and duties

○   The employer must not hinder or prevent the contractor from carrying out and completing the works as envisaged by the contract. Breach of this duty will render the employer liable to an action for damages and may disentitle him from enforcing the liquidated damages clause.
○   Should the contractor fail to complete the works by the completion date, the employer is entitled to recover liquidated damages.
○   The employer must pay the contractor on your certificates, payment being due within 14 days of the date of each certificate.
○   The retention fund may be used by the employer to satisfy bona fide and quantified claims but it is the contractor's money and not the employer's.

Chapter 7

Quantity Surveyor and Clerk of Works

7.1   Quantity surveyor

7.1.1 Appointment

MW 80 makes provision for the appointment of a quantity surveyor in Recital 4 which contains the only reference to the quantity surveyor in the entire contract. There is a footnote which directs you to 'Delete as appropriate', i.e., if no quantity surveyor is appointed.

The decision whether or not to appoint a quantity surveyor will be taken by the employer with your advice. Your advice will naturally take into account the amount of work which the quantity surveyor could be asked to carry out. In the case of small works, the use of a quantity surveyor tends to be the exception. You should be competent to deal with interim payments, the computation of the final sum and valuation of variations if the work is of a simple and straightforward nature, particularly if you insert a clause to provide for stage payments. If, however, you have any doubts about your competence in this field or the work is complex, the use of a quantity surveyor is indicated. It is important to remember that, if you hold yourself out to your client as capable of carrying out quantity surveying functions, that is the standard which will be expected of you. Do not fill in your own name as quantity surveyor unless you have checked your position with your professional indemnity insurers. Since there is no further mention of the quantity surveyor in the contract, there appears to be no valid reason why you should ever insert your own name. There is no benefit in so doing, and there may be insurance problems

If you decide that a quantity surveyor is required for the works, you should inform the employer at the earliest possible opportunity, usually stage A – Inception, so that he can assist you throughout the project. Take care to explain to the employer why a quantity surveyor is required (Fig

7.1). The situation will be simplified if the employer has been given a copy of *Architect's Appointment* on which, presumably, your engagement is based.

7.1.2 Duties

The contract goes no further than to provide for the appointment of a quantity surveyor; it is silent regarding his duties. Although it is probable that an implied term would govern the matter, if a quantity surveyor is appointed it is best to deal with his powers and duties by means of a suitably worded provision which gives him a right of access to the site. There are two possible ways of dealing with this:

o   Insert a brief clause in the contact to cover the activities you intend the quantity surveyor to perform.

o   Write to the contractor informing him that the quantity surveyor is to be your authorised representative in respect of specified activities.

The introduction of a clause in the contract is probably the more satisfactory way of accomplishing your objective. The contractor then knows, at tender stage, just what is intended. If you handle the matter by means of a letter at the beginning of the contract, the contractor might object violently to it and you will have soured relations at the start. Another danger is that the contractor may not appreciate the limits of the quantity surveyor's duties and carry out as instructions what are simply the quantity surveyor's comments on some aspect of the work. Although the contractor would be wrong to do so, it is little consolation to you or the employer if disruption and delay results. If you feel that you must deal with the quantity surveyor's duties in this way, take great care with your letter (Fig 7.2).

If you include a clause, get proper legal assistance. Among the duties that you wish the quantity surveyor to carry may be the following:

o   Value the amount of the employer's contribution under clause 2.5, if you instruct otherwise than that the contractor should make good all defects at his own expense (see section 9.3.4).

o   Valuation work under clause 3.5 if the employer has to employ others to carry out work contained in your instructions.

o   Valuation of variations in accordance with clause 3.6.

o   Measurement and valuations under clause 4.2.

o   Measurement and valuations under clause 4.3.

o   Computation of the final sum under clause 4.4.

o   Calculations under clause 4.5.

o   Calculations under clauses 6.3A and 6.3B.

o   Valuations under clause 7.1 and 7.2.

Remember to include provision for the quantity surveyor to require the contractor to supply any necessary documents for the purpose of carrying out the valuations, calculations or measurements. It is also prudent to

Fig 7.1

Letter from architect to employer: if the services of a
Quantity Surveyor are required

PROJECT TITLE

Dear Sir

In view of the particular nature/size/value [*delete
as appropriate*] of this project, the services of a
quantity surveyor will be required. He is the
specialist in building economics and he will deal
with the preparation of valuations for progress
payments and variations and provide overall cost
advice. He should be appointed at this stage so that
you can derive the maximum benefit from his advice
and he can become fully involved with the project.

I suggest the use of [*insert name*] of [*insert
address*], with whom I have worked many times in the
past. If you will let me have your agreement, I will
carry out some preliminary negotiations on your
behalf and advise you regarding the letter of
appointment.

The use of consultants is covered by clauses 2.45 and
3.5 to 3.7 inclusive of "Architect's Appointment", a
copy of which is already in your possession.

Yours faithfully

Fig 7.2

Letter from architect to contractor: regarding the duties of
the Quantity Surveyor

PROJECT TITLE

Dear Sir

The quantity surveyor named in Recital 4 of the
contract is [*insert name*] of [*insert address*]. This
letter is to notify you that he will be my authorised
representative to carry out the duties of valuation,
calculation, computation and measurement in respect
of the following clauses of the Conditions of
Contract:

2.5

3.5

3.6

4.2

4.3

4.4

4.5

6.3A/6.3B [*delete as appropriate*]

7.1

7.2

You must supply him with all necessary documents,
vouchers, etc, to enable him to carry out his work.
Your attention is drawn to the fact that the quantity
surveyor's duties are limited to quantification. He
is not empowered to issue instructions, certificates,
make awards or decide liability for payment.

Yours faithfully

Copies: Employer
        Quantity surveyor

stress that the quantity surveyor's duties cover quantification only with no power to agree or decide liability for payment nor to issue instructions, certification or make any awards, though this would be the situation under the general law.

7.1.3 Responsibilities

The quantity surveyor's responsibility is to the employer in contract and in tort. Even if you include a list of his duties in a special clause inserted in MW 80, he will not be liable to the contractor in contract because, like you, he is not a party to the contract. In limited circumstances he might be liable to the contractor in tort but you are the professional entrusted with the task of certifying all payments. If the quantity surveyor makes a mistake, you must correct it. The onus on you to check all quantity surveyor's calculations cannot be emphasised too much.

Clause 3.6 of *Architect's Appointment* states that, where your client employs a consultant, he will hold that consultant responsible for the competence, general inspection and performance of the work entrusted to that consultant. You should not place too much reliance on the wording of the clause, however, if the quantity surveyor makes a mistake, because you will always be the first target for your client's displeasure. Moreover, the clause goes on to state that nothing in the clause will affect your own responsibility for issuing instructions or for other functions ascribed to you under the contract. One of the 'other functions', of course, is the certification of payments to the contractor.

If the quantity surveyor gives advice directly to the employer regarding any matter, he is liable if the advice is negligently given. In practice, this will happen only rarely. An example might be if the quantity surveyor is present at the meeting to open tenders for the main contract and volunteers negligent advice, as a direct result of which the employer enters into a contract with a contractor which turns out to be more expensive than the employer was led to believe. Even in this sort of instance, there is a very good chance that the employer will blame you for the difficulty. The employer may have a point, but everything will turn on the precise circumstances in which the quantity surveyor gave his advice. The quantity surveyor has exactly the same sort of contractual relationship with the employer as you have. Your duties are different and you have, in addition, the authority to co-ordinate and integrate the consultant's services (*Architect's Appointment*, clause 3.7), but you can never be responsible for the quantity surveyor's actions or defaults.

You should resist, as far as possible, any efforts on the part of the employer to get you to appoint the quantity surveyor yourself. From his point of view it is understandable that he wishes to deal with everything through one person, but the arrangement leaves you very exposed. The modern practice in negligence cases seems to be to sue everyone in sight and some commentators argue that it makes no real difference whether the employer appoints all consultants directly or simply appoints the architect and leaves him to appoint any other consultants, with the employer's permission, who may be necessary. We do not share that view. Fig 7.3 indicates the various relationships which are possible. If the employer suggests that you appoint the quantity surveyor, we suggest that you put your position on record (Fig 7.4).

Remember that you have no contractual relationship with the quantity surveyor (if the employer has appointed him); any liability between you will be in tort.

7.2   Clerk of works

7.2.1 Appointment

MW 80 makes no provision for a clerk of works. This is because, quite clearly, on most works for which MW 80 is suitable, the employment of a clerk of works in full- or part-time capacity is hardly justified. Every project has its own difficulties, however, and if you consider that a clerk of works is required, you must advise the employer accordingly.

The *Architect's Appointment* expressly states in clause 3.10 that you will not be required to make frequent or constant inspections. Clause 3.11 continues, to state that where such inspections are required, a clerk of works will be employed. If the employer is one of those organisations which employ clerks of works on their permanent staffs, that is an excellent arrangement. To avoid misunderstandings, be sure to put the position on record (Fig 7.5). In the case of most employers, however, it will be a one-off arrangement, the clerk of works being engaged by the employer on a full-time or part-time basis for the duration of the contract. Some firms of architects employ their own clerks of works on a permanent basis, but, in the light of recent case law, you are advised to avoid employing a clerk of works yourself, even though the employer pays you an additional sum to cover his salary (see section 7.2.3).

If a clerk of works is to be used, it is essential to include a clause to that effect in the contract. It is suggested that the clause follows the lines of clause 3.10 of IFC 84 Intermediate Form which simply states:

'The Employer shall be entitled to appoint a clerk of works whose duty shall be to act solely as an inspector on behalf of the Employer under the directions of the Architect.'

This states all that is necessary and removes any uncertainty in the contractor's mind regarding the directions of the clerk of works – he is not empowered to issue any.

In the interests of the project, you should advise the employer to appoint

Fig 7.3

Possible contractual relationships

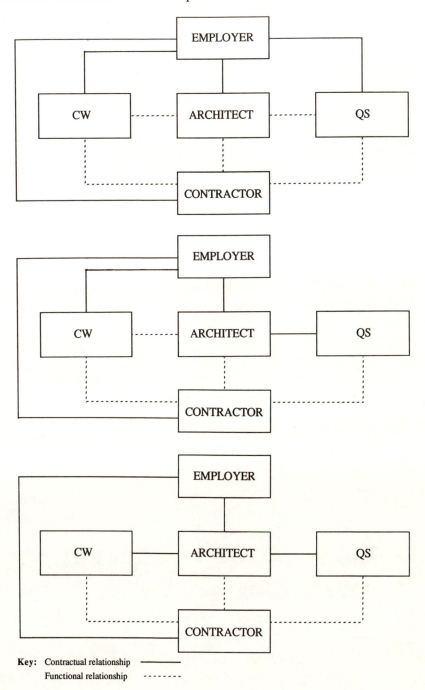

Key: Contractual relationship ————
     Functional relationship - - - - - - -

Fig 7.4

Letter from architect to employer: if he requires you to appoint the quantity surveyor

PROJECT TITLE

Dear Sir

Thank you for your letter of the [*insert date*] and I am pleased that you have agreed to the appointment of [*insert name*] as quantity surveyor for this project.

I strongly advise you to appoint the quantity surveyor directly yourself. That is normal practice in construction projects and gives you direct right of access to him if you should so wish. I cannot provide these services myself and, if you wish me to appoint, I should do so simply as your agent (see "Architect's Appointment", clause 3.6).

I will, in any case, co-ordinate all professional work. The quantity surveyor's fees are the same whether appointed directly, as I advise, or through my office. If you let me know that you will take my advice, I will draft an appropriate letter of appointment for you to send.

Yours faithfully

Fig 7.5
Letter from architect to employer: regarding the appointment of a clerk of works


```
PROJECT TITLE

Dear Sir

In view of the nature/size/value [delete as
appropriate] of this project, I advise the
appointment of a clerk of works on a part-time basis,
say [insert number] hours per week.

The "Architect's Appointment", clause 3.10, provides
that I shall not be required to make frequent or
constant inspections to check the quality of the
work.  A clerk of works will be able to make
inspections of such frequency as should ensure the
proper carrying out of the work.  He is likely to
repay his cost several times over in savings on lost
time and money as the contract progresses.

Although naturally, I will carry out my own duties
with proper skill and care, on a contract of this
type I cannot accept responsibility for such defects
as would be discovered by the employment of a clerk
of works.

Yours faithfully
```

Fig 7.6

Letter from architect to clerk of works: setting out duties

PROJECT TITLE

Dear Sir

My client [*insert name*] has confirmed your appointment as clerk of works for the above contract. I should be pleased if you would call at this office on [*insert date*] at [*insert time*] to be briefed on the project and to collect your copies of drawings, specification, schedules, weekly report forms and site diary.

The contractor is expected to take possession of the site on the [*insert date*]. You will be expected to be present on site for a minimum of [*insert number*] hours per week from that date. I will discuss the timing of your visits when we meet. Let me know at the end of the first week if proper accommodation is not provided for you as described in the specification.

Your duties will be as indicated in the Conditions of Contract special clause [*insert number*], a copy of which is enclosed for your reference. In particular, I wish to draw your attention to the following:

1. You will be expected to inspect all workmanship and materials to ensure conformity with the contract, i.e. the drawings, specification, schedules and further instructions issued from this office. Any defects must be pointed out to the person in charge, to whom you should address all comments. If any defects are left unremedied for twenty-four hours or if they are of a major or fundamental nature, you must let me know immediately by telephone.

2. Although it is common practice for clerks of works
    to mark defective work on site, you must not make
    such marks or in any way deface materials on site.

3. It is not my policy to issue lists of defects to
    the contractor before practical completion
    (commonly known as "snagging lists"). They are
    open to misinterpretation and should be compiled
    by the person in charge. Confine yourself to oral
    comments.

4. The architect is the only person empowered to
    issue instructions to the contractor.

5. Any queries, unless of a minor explanatory nature,
    should be referred to me for a decision. You are
    not empowered to vary or omit work.

6. The report sheets must be filled in completely and
    a copy sent to me on Monday of each week. Pay
    particular attention to listing all visitors to
    site and commenting on work done in as much detail
    as possible.

7. The diary is provided for you to enter up your
    daily comments.

8. Remember that your weekly reports and site diary
    may be called in evidence should a dispute arise
    so you must bear this in mind when making your
    entries.

The successful completion of the contract depends in
large measure upon your relationship with the
contractor. If you are in any doubt about anything,
please let me know.

Yours faithfully

the clerk of works as soon as the contractor's tender has been accepted. This gives the clerk of works time to thoroughly familiarise himself with the drawings, specification and schedules. You will hold a meeting with him to brief him about the work to be done, but it is wise to confirm the main points to the clerk of works by letter (Fig 7.6) which should outline the aspects of his job you consider to be important and remind him of the extent and limitations of his duties.

7.2.2 Duties

If you incorporate a clause based upon IFC 84 clause 3.10, the clerk of works' sole duty is to inspect the works. The clause makes clear that he carries out this function on behalf of the employer; he is not your inspector. This is vitally important as will be seen in the next section. The employer, however, has no power to direct him. That is your prerogative alone.

Your directions to the clerk of works will presumably embrace how, where and at what intervals he should inspect and to what he should pay particular attention. If you issue any directions which are other than purely routine in nature, always confirm them in writing to protect your position. In practice, the clerk of works will often do more than simply inspect. He will usually be a person of some experience whose advice you will seek from time to time. The contractor often asks the clerk of works for assistance in solving site problems. Note, however, that, if the contractor acts on the advice or even instructions of the clerk of works, he does so at his peril. In giving advice, the clerk of works is acting in a personal capacity. Make sure that the contractor understands the position at the beginning of the contract. It is prudent to give some time to this topic at the first contract meeting and give some space to it in the minutes. Contractors get into the habit of accepting the clerk of works as speaking on your behalf. Although there is no foundation for this view in the contract, it saves much bad feeling to put the matter beyond doubt. One of the worst things that can happen on a contract is for you to have to overrule the clerk of works.

The duties of the clerk of works can, thus, be seen in two parts:

o   His duties under the contract.
o   His duties by virtue of your specific directions to him.

His duties under the contract define his relations with the contractor and his duty to comply with your directions defines his relations with you. That is to put the matter in broad terms. Every architect has his own views about his relationship with the clerk of works, but among the duties you might expect him to fulfill are the following:

o   Inspect the works.
o   Relay queries and problems back to you.

o   Complete report sheets.

o   Complete daily diary.

o   Take measurements as directed.

o   Take particular notes of such things as portions of the work opened up for inspection under clause 3.5.

The clerk of works is not empowered to put any marks on defective portions of the work. Once such work is removed, it is the property of the contractor who is entitled to expect that the clerk of works will not do any damage to the works, no matter how slight.

It is, unfortunately, common practice for the clerk of works to issue 'snagging lists' to the contractor, particularly towards the end of a job. You should discourage the practice because:

o   The clerk of works is inspecting for the benefit of the employer, he owes no duty to the contractor to find defects.

o   It is the job of the person in charge to produce his own lists of work.

o   The contractor may be under the impression, however misguided, that his obligation relates only to the 'snagging lists' and disputes may follow.

Obviously, the clerk of works must draw the contractor's attention to work not in accordance with the contract documents, but he should not be more specific. In particular, neither he nor you should instruct how defective work is to be corrected if the only problem is that it is not in accordance with the contract.

7.2.3 Responsibilities

Like everyone else connected with the contract, the clerk of works has a responsibility to carry out his duties in a competent manner. He must demonstrate the same degree of skill that would be demonstrated by the average clerk of works. If he holds himself out to be especially skilled in some branch of his work, a greater standard of skill than the average will be expected in that particular branch.

Recent case law has made clear that, provided the ordinary relationship of master and servant exists between employer and clerk of works, the employer will be vicariously liable for the actions of the clerk of works in the normal way: *Kensington and Chelsea and Westminster Area Health Authority* v *Wettern Composites & Others* (1984) 1 ConLR 114. The fact that the clerk of works is under your direction makes no difference. The relationship between the clerk of works and yourself has been compared to that between the Chief Petty Officer and the Captain of the Ship. This does not mean that, if the clerk of works is negligent, it will relieve you of all responsibility, but it may substantially reduce your liability for damages depending upon circumstances. It is, therefore, very much in your interests to ensure that the employer employs a clerk of works. If *you* employ the clerk of works, his negligence will not reduce your own liability.

7.3   Summary

The quantity surveyor

o   The contract provides for the appointment of a quantity surveyor, but not for his duties.

o   It is for you to advise the employer on the appointment.

o   There is danger in holding yourself out as capable of carrying out quantity surveying functions.

o   If you act as quantity surveyor, check with your professional indemnity insurers.

o   You should insert a clause in the contract or write a letter to the contractor to cover his duties.

o   His duties should include deciding quantum, but not liability to pay.

o   The quantity surveyor should be appointed by the employer, to whom he will be liable in contract.

o   It is your responsibility to check that all certificates are correct.

o   Despite legal liabilities, the employer will look to you first if something goes wrong.

Ther clerk of works

o   No provision in the contract.

o   Advise the employer if you think a clerk of works is necessary.

o   Insert a suitable clause in the contract.

o   The clerk of works should be appointed by the employer.

o   Should be appointed immediately the successful tender has been accepted.

o   Brief him thoroughly.

o   Sole duty should be to inspect the works.

o   He may carry out other duties for you.

o   He must not put marks on the work.

o   He should not issue 'snagging lists'.

o   He should not instruct how defective work is to be made good in accordance with the contract.

o   He must be competent.

o   His employment may reduce your liability for damages if the employer is found to be vicariously liable.

Chapter 8

Sub-Contractors and Suppliers

8.1   General

This chapter deals with third parties in so far as they are provided for or might affect the work carried out under MW 80. Sub-contractors, suppliers, statutory authorities, persons engaged directly by the employer and the possibility of nominating or naming sub-contractors are also considered.

The contract provisions are extremely brief. They are contained in clauses 3.1, 3.2 and 5.1. There are no specific provisions for suppliers, employers, employer's licensees or nominated or named sub-contractors.

8.2   Sub-contractors

8.2.1 Assignment

Assignment is usually coupled with sub-contracting in contract provisions and MW 80 is no exception. It is a mistake, however, to consider that they are linked. They are totally different concepts. Assignment is the legal transfer of a right or duty from one party to another whereby the original party retains no interest in the right or duty thereafter. For this to be fully effective *novation* must take place, that is, the formation of a new contract.

Sub-contracting is, in essence, the delegation of a duty from one party to another, but the original party still retains primary responsibility for the discharge of that duty. It is vicarious performance of a duty by someone else.

Clause 3.1 deals with assignment. Both employer and contractor are prohibited from assigning the contract unless one has the written consent of the other. This is a much stricter provision than is to be found under the general law where it is possible for either party to assign the benefits or rights of a contract to a third party. For example it is quite common for the contractor to wish to assign to a third party the benefit of receiving progress payments in return for substantial financial help at the beginning of the contract. The employer, too, may wish to sell his interest in the completed building to another before the issue of the final certificate, thus assigning the benefit of the contract. Although permitted under the general law, clause 3.1 operates to stop such deals. Duties or burdens of contracts can never be assigned without express agreement between the parties.

It is considered that you have a duty to explain the possible difficulties to the employer, particularly if you have reason to believe that he might wish to assign his benefit before the final certificate is issued. Assignment can be made with the written consent of the other party, but such consent may be refused and the grounds for refusal need not be reasonable. It would appear to do no harm, and make much sense, to advise the employer to amend the contract so as to prohibit only the assignment of duties under the contract. The amendment should be carried out at tender stage and should result in lower rather than higher tenders.

8.2.2 Sub-contracting

Clause 3.2 deals with sub-contracting. Sub-contracting is traditional in the building industry, but the practice can be abused to the extent that the contractor's sole employee on the site might be the person-in-charge, the remainder of the workforce being sub-contractors. Needless to say, such an arrangement does not make for an efficient contract. Clause 3.2 is designed to prevent this and other problem situations by allowing sub-contracting only if you give your written consent. Although you may withhold your consent, you must act reasonably. It is probably reasonable for you to require the contractor to supply you with the names of sub-contractors before you consent.

It is important to remember that there is no contractual relationship between the employer and the sub-contractor. The sub-contractor's contract is with the contractor. It follows that nothing contained in MW 80 is binding in any way upon the sub-contractor. It is vital, therefore, that the sub-contract gives the contractor sufficient control over the sub-contractor because the employer must look to the contractor for redress if the sub-contractor defaults. It is thought that you would not be unreasonable if you withheld your consent to sub-letting until you had satisfied yourself that the contractor's sub-contract provisions were adequate. Beware, however, of being drawn into disputes between contractor and sub-contractor or of being seen to approve the form of sub-contract. Remember that ultimately the contractor is responsible for his own sub-contractors and he has no right to look to you or the employer to assist him if things go wrong.

The lack of detailed provisions is something to be regretted, but unavoidable taking the contract as a whole. Unfortunately, although the number of sub-contractors will tend to increase with the size of the work, there is no lower point at which the contractor will not use any sub-contractors at all. MW 80 makes no reference to the terms which the contractor must include in the sub-contract. Although such a list would not be binding on the sub-contractor, it would be binding on the main contractor and, probably of more importance, it would form a valuable checklist of matters to be included. If you do decide to check the contractor's form of sub-contract and it is not a recognised form, Table 8.1 shows some of the matters for which it should make provision. Remember, you have no responsibility for the sub-contract and if you check it, you will do so merely to ensure that the contractor has avoided situations which might have repercussions on the main contract. Any dispute between contractor and sub-contractor is potentially disruptive.

8.2.3 Nominated sub-contractors

MW 80 makes no provision for nominated sub-contractors. Their use for works for which this form would be suitable is not envisaged. Having said that, there are various devices which can be used to provide for 'nominated' sub-contractors if you or the employer are absolutely sure that you must 'nominate' a particular firm.

Table 8.1
Sub-contract checklist

- Sub-contractor's obligations.
- Information: supply and timing of documents, drawings and details; custody and confidentiality; design liability.
- Instructions, variations and valuations.
- Access to site for sub-contractor; access to sub-contract works for architect; regulations.
- Assignment and sub-contracting.
- Vesting of property and insurance.
- Commencement, completion and extension of the sub-contract period.
- Completion date and making good of defects.
- Payment.
- Financial claims.
- Determination.
- Finance (No. 2) Act, 1975.
- Disputes procedure.

The widespread use of nominations in building contracts has come to be associated with a certain laziness on the part of the architect. It often seems quicker and less trouble in the short term to put a sum in the bills of quantities or specification than to properly specify requirements at the outset. It is obviously a temporary expedient only and at some future date, usually sooner rather than later, the architect is faced with specifying the work in question and engaging in the complicated process of nomination. Difficult legal and administrative problems can result during the life of the contract – as numerous legal cases testify. The message is clear. Wherever possible, avoid nomination. If you are still intent on nominating, you should consider using JCT 80, which has detailed (but complex) provisions, or IFC 84, which provides for 'naming' sub-contractors in a clause of equal complexity and some obscurity, or ACA 2, which has relatively simple 'naming' procedures. If you wish to nominate using MW 80, it is possible to use one of the following methods:

o   Name one or a choice from several named firms in the specification.
o   Name a firm in your instruction directing the expenditure of a provisional sum in accordance with clause 3.7.
o   Include an appropriately worded clause in the contract.
o   Provide for the specialist firm to be directly employed by the employer.

There are severe pitfalls inherent in the use of any of these methods and you should seek the advice of an experienced building contracts specialist before you proceed. Among the points to be considered are:

Naming in the specification
It is perfectly possible for you to name one firm in the specification which the contractor must use for carrying out a specific part of the work. If you do this, you place a considerable amount of power in the hands of such a firm which can, in effect, quote to the contractor whatever price it likes secure in the knowledge that whichever contractor is awarded the contract, the specialist firm will be incorporated. If, as is not unknown, the specialist firm goes into liquidation before work begins on site, you have an instant problem with which to start your contractual duties.

The biggest difficulty with this type of nomination is that there is no provision in the contract to deal with the consequences. The precise extent of those consequences can be envisaged by glancing through clause 35 of JCT 80. The answer, of course, is to incorporate a fairly substantial clause in the contract to deal with the situation. Such a clause would require very careful drafting.

The alternative, whereby you give the contractor a choice of three or four names in the specification, is not strictly nomination at all. However, it does give you a degree of control over the firms to be used, while allowing the contractor to seek competitive quotations. A sub-contractor appointed by the contractor under this method would be the contractor's entire responsibility to the extent that, if he failed, it would be the contractor's job to find an alternative. Although the worst consequences of nomination can be avoided by this system, it is advisable to include a suitably worded clause in the contract.

Naming by an instruction in respect of a provisional sum
If you intend to operate this system, it is essential that the contractor knows what you intend to do at the time of tender so that he can make suitable provision in his price. A big danger is that the contractor might strongly object to the firm in question when you name it for the first time in your instruction. Although there is no machinery to deal with his objection, the efficient progress of the work can be seriously affected. All the comments regarding the naming of a single firm in the specification are equally applicable to this case and an amendment to clause 3.2 and 3.7 is indicated.

Including a clause in the contract
The trouble with this is that once you start to add substantial clauses dealing with particular circumstances, in this case nomination, you run the risk of the contract being considered the employer's 'standard written terms of business' which brings it within the scope of the Unfair Contract Terms Act 1977. There is no doubt that a suitable clause can be formulated and, if nomination is what you want, it is probably the safest way to achieve it. Nomination can proceed along time-honoured procedures and potential difficulties can be provided for.

The employer directly employing the specialist firm
This may seem an attractive solution, but there are dangers, not least the problem of integration with the works (see section 8.4).

8.3 Statutory authorities

MW 80 does not expressly mention statutory authorities. However, clause 5.1 states that the contractor must comply with any statute, any statutory instrument, rule or order or any regulation or byelaw applicable to the works. This means, for example, that he must not contravene the Planning Acts or regulations made in pursuance thereof and he must comply with the Building Regulations. By necessary implication, therefore, he must allow statutory authorities to enter upon the site and carry out work which they alone are empowered to do.
Certain crucial parts of virtually all contracts are carried out by statutory authorities such as local authorities, gas, water and electricity boards. When they carry out work solely as a result of their statutory rights or

obligations, they are in a special category quite separate from sub-contractors or employer's directly employed firms. If completion of the works is delayed as a result of an authority carrying out or failing to carry out work in pursuance of its statutory duty, the contractor will be entitled to an extension of time in accordance with clause 2.2.

If the contractor employs a statutory undertaking to carry out work which is not part of its statutory rights or duties, the statutory undertaking ranks as an ordinary sub-contractor. Delay caused by the undertaking in such a case would not entitle the contractor to an extension of time under MW 80. An example should make the principle clear.

It is usual to include a provisional sum in the specification to cover the cost of connecting the electrical system of a dwelling to the mains. The contractor may also, with your written permission, sub-let the electrical wiring and fittings in the dwelling to the electricity board. The mains connection is part of the board's statutory obligations; the internal wiring and fittings is not part of the board's statutory obligations, but is a matter of contract between the contractor and the board. If the mains connection delays the completion date, the contractor is entitled to an extension of time. If the internal wiring and fittings delay the completion date, the contractor is not entitled to any extension of time.

Statutory authorities have no contractual liability when they are carrying out their statutory duties – a sore point with many architects – but in certain cases they have a liability in tort. Outside their statutory duties, they are in the same position as anyone else if they enter into a contract to carry out work.

The contractor must also give all notices required by statute, etc., and pay all fees and charges in respect of the works provided that they are legally recoverable from him, but not otherwise. He is not entitled to be reimbursed and he is deemed to have included the necessary amounts in his price. The exception to this, of course, is if the charge is a necessary result of an instruction you have given. In such a case, the amount of the charge will form part of the valuation of your instruction carried out under the provisions of clause 3.6.

Clause 5.1 states that the contractor is not liable to the employer under the contract if the works do not comply with statutory requirements provided that:

○ He has carried out the works in accordance with the contract documents or any of your instructions *and*

○ If he has found a divergence between the contract documents or your instructions and statutory requirements, he has given you immediately a written notice specifying the divergence.

The contractor is not liable if he fails to find a divergence which actually exists: *London Borough of Merton* v *Stanley Hugh Leach Ltd* (1985) unreported. Although the contractor may be freed from liability to the employer, his duty to comply with statutory requirements remains. Thus, the local authority may serve notice on him if work, built correctly in accordance with the contract documents, does not comply with the Building Regulations. In such a case, you would have to act speedily to issue appropriate instructions if the employer was to avoid a substantial claim at common law for damages.

The contractor is not entitled to take any emergency measures to comply with statutory requirements even though delay might cost the employer money. His obligation to give you immediate written notification remains. If the emergency concerns part of the structure which is actually dangerous, the contractor has a responsibility under the general law to take whatever measures are necessary to make the structure safe. It is difficult to see how he could make any valid claim in such circumstances.

8.4 Works not forming part of the contract

The contract makes no provision for the employer to enter into a contract with anyone other than the contractor to carry out any part of the work on site. It is quite common for the employer to wish to engage others to do certain work or, indeed, use some of his own employees. The reason may be because the employer has a special relationship with the firm or individual, for example in the case of a sculptor, artist or landscaper, or because the employer wants complete control over a particular operation. When using MW 80, direct engagement by the employer can be achieved in one of two ways:

o   With the consent of the contractor.

o   By including a special clause to that effect.

It is never a good idea to bring third parties onto a site during the contract period. Three dangers which merit special consideration are:

o   It is easy for the contractor to claim that such persons have disrupted his work and/or delayed the completion date. They are very difficult claims to refute because introducing third parties clearly does not help the contractor and can usually be seen to cause at least some degree of hindrance. The contractor may be able to claim damages at common law and an extension of time under clause 2.2

o   The contractor may acquire grounds to determine his employment under the contract. Clause 7.2.2 provides for the contractor to determine if 'the employer or any person for whom he is responsible interferes with or obstructs the carrying out of the Works . . .' The matter is dealt with in more detail in section 12.3.2

o   For insurance purposes, persons directly employed by the employer are deemed to be persons 'for whom the employer is responsible' (clause 6.1). They are not deemed to be sub-contractors. Therefore the employer may have uninsured liabilities. The directly employed persons may well have their own insurance cover, but it is your duty to advise the employer to obtain the necessary cover through his own broker. The cover should be for the employer and those persons for whom he is responsible in respect of acts or defaults occurring during the course of the work. It is a complex business and best left in the broker's hands.

Remember that statutory authorities, acting outside the confines of their statutory duties, may be considered to be directly employed by the employer if they are not paid by the contractor and under his control.

It is always prudent to ensure that the contractor is responsible for all the work to be carried out during the currency of a contract. It removes some possible areas of dispute and promotes efficiency. If the employer insists on having directly employed persons, remember that, for work to be considered as not forming part of the contract, it must:

○ Be the subject of a separate contract between the employer and the person who is to provide the work *and*

○ Be paid for by the employer direct to the person employed, not through the contractor.

8.5 Summary

Assignment

○ A different concept from sub-contracting.
○ Neither party may assign without the other's consent.
○ Under the general law, either party may assign benefits, but not duties.
○ It may be beneficial to amend clause 3.1.

Sub-contracting

○ The contractor may sub-let with your consent.
○ There is no contractual relationship between the employer and the sub-contractor.
○ The sub-contract terms may have important repercussions for the employer.

Nominated sub-contractors

○ No provision in MW 80.
○ If nominated sub-contractors are required, consider using a different contract form.
○ Devices can be used to enable nominations to be made when using MW 80, but there are pitfalls.

Statutory authorities

○   The contractor must allow them to enter the site.
○   He must comply with statutory requirements.
○   In pursuance of their statutory duties, they are not liable in contract, but may provide grounds for an extension of time.
○   Not in pursuance of their statutory duties, they are liable in contract and may be sub-contractors or persons for whom the employer is responsible.
○   The contractor is not liable to the employer if he works to the contract documents, provided that he notifies you of any divergence he finds.
○   There is no provision for emergency work.

Work not forming part of the contract

○   MW 80 makes no provision for directly employed persons.
○   The contractor must consent or you must include a special clause.
○   Fertile ground for claims.
○   Possible ground for determination.
○   Insurance implications.
○   Should be avoided.

Chapter 9

Possession, Completion and Defects Liability

9.1 Possession

9.1.1 Introduction

If there is no express term in the building contract, a term will always be implied that the contractor must have possession in sufficient time to allow him to finish the works by the contract completion date: *Freeman* v *Hensler* (1902) 2 HBC 231. To have possession of something is the next best thing to ownership. If you lend this book to a friend, he can defend his claim to it against anyone except you. The same principle applies to a builder in exclusive possession of a site. He is in control of the site and has the power to refuse access to anyone else, including the employer. In practice, this stern rule is modified by the operation of numerous statutory regulations, allowing entry by the representatives of various statutory bodies, and by the express and implied terms of the contract.

In legal terms, the contractor is said to have a licence from the owner of the site to occupy it for the period of time necessary to carry out and complete the works. The period of time is the period stated in the contract or any extended period. During the contractor's lawful occupation the employer has no power under the general law to revoke the licence, but the contract may contain express terms giving such power, for example, in the case of lawful determination of the contractor's employment. Case law suggests that the absence of such an express term can pose awkward problems if the contractor refuses to give up possession. In general, however, if the contractor retains possession of the site after the contract period or any extended period has expired, he is in the position of a trespasser and can be removed under the general law.

This contract does contain an express term requiring the contractor to give up possession immediately if the employer determines the contractor's employment under the provisions of clause 7.1. Although there is no similar specific requirement when the contractor determines his own employment under clause 7.2, a term to that effect will be implied.

There is no express provision for access to the works for the architect equivalent to clause 11 of JCT 80, but such a term must be implied to allow the architect and his authorised representatives to carry out their duties under the contract.

9.1.2 Date for possession

JCT 80, IFC 84, ACA 2 and GC/Works/1 have an Appendix, Time Schedule or Abstract of Particulars in which important dates, sums of money and other variables are to be filled in by the parties before the contract is signed. The Appendix becomes a vital contract document. MW 80 has no such Appendix, but important dates, etc., are to be inserted in appropriate places in the clauses themselves.

Clause 2.1 relates to 'Commencement and completion', but it does not make reference to a date for possession. It does, however, leave a space for a date to be inserted on which the works 'may be commenced'. It will be implied that this date is the latest date on which the employer must give possession of the site to the contractor. The inclusion of the word 'may' could be significant. The straightforward interpretation of the clause appears to be that the date to be inserted is the earliest on which the contractor will be allowed to commence carrying out the works, but not necessarily the latest. In effect, it appears that the contractor would be within this clause if he started work in the fourth month of a six month contract, provided that he finished by the contract completion date. Contrast with JCT 80, clause 23.1, where the contractor 'shall thereupon begin the Works . . .' Of course, in practice, the contractor could well be in danger of having his employment determined in accordance with clause 7.1.1 – 'if the Contractor without reasonable cause fails to proceed diligently with the Works or wholly suspends the carrying out of the Works before completion' – he would not be proceeding diligently, but it is a moot point whether he could be said to be suspending something he had not yet begun. In any case, he might well argue that he had reasonable cause if, in fact, the contract period was very generous.

If the employer fails to give the contractor sufficient possession for him to begin the works on the stated date, it will be a serious breach of contract. The contractor will have a claim for damages at common law and the completion date may well cease to be operative. In such circumstances the contractor's obligation would be to complete within a reasonable time. Although that does not mean that the contractor has unlimited time in which to carry out the works, it does mean that there is no date from which liquidated damages can run and, therefore, they are not deductible. If the employer's failure to give possession lasts more than a few days, it is our view that the contractor may well have grounds to consider the employer's breach as an intention to repudiate the contract. In any case, failure to give possession clearly requires negotiation between the employer and the contractor to achieve an amicable settlement and you should remind the employer of his duties before the due date (Fig 9.1) and give him your advice if he is in breach (Fig 9.2).

Although your power to instruct the contractor certainly extends to postponing the work, postponement is not the same as failure to give possession. If you postpone the work, the contractor will have possession of the site, but the carrying out of the work will be suspended. The contractor may well wish to use the time to reorganise his site arrangements, repair site offices, improve his security, etc.

When the contractor has completed the work, he normally gives up possession, but he has a restricted licence to continue to enter upon the site to deal with such defects as are notified to him under clause 2.5, Defects liability (see section 9.3).

9.2   Practical completion

9.2.1   Definition

Clause 2.1 states that the works must 'be completed by' a date to be inserted. The words have their ordinary meaning, that is to say the completion of the works must not take place after the stated date, but it may take place before the stated date. Note, however, that there is no provision for partial possession by the employer. This omission is entirely reasonable in view of the small-scale nature of the works for which this contract is intended to be used and the correspondingly short time scale. It is extremely unlikely that the employer will need to take possession of part of the work before completion. If it is decided that provision must be made for partial possession, it is worth considering the use of IFC 84 instead, incorporating the provision for partial possession detailed in Practice Note IN/1, page 10. If phased completion is to be provided, neither MW 80 nor IFC 84 is suitable and either JCT 80 with the Sectional Completion Supplement or ACA 2 would be appropriate. It is not envisaged that partial possession, much less sectional completion, would be a feature of work for which MW 80 was being considered.

Clause 2.4 states that you must 'certify the date' when in your opinion the works have reached practical completion. In accordance with clause 1.2, you must issue the certificate in writing. Although no time scale is indicated, you must issue the certificate within a reasonable time of the certified date because it is a particularly important stage in the contract

Fig 9.1
Letter from architect to employer: before date for possession

PROJECT TITLE

Dear Sir

The contractor is entitled, by the terms of the contract, to take possession of the site on the [*insert date*].

Will you be certain that everything is ready so that the contractor can take possession? Failure to give possession on the due date is a serious breach of contract which cannot be remedied by a simple extension of the contract period, the contractor may be able to claim substantial damages or even treat the contract as repudiated.

Please let me know immediately if you anticipate any difficulties.

Yours faithfully

Fig 9.2
Letter from architect to employer: if he fails to give possession on the due date

PROJECT TITLE

Dear Sir

I understand/have been notified by the contractor
[*delete as appropriate*] that you were unable to give
him possession of the site on the date stated in the
contract as the date on which he may commence the
works.

You will recall that in my letter of the [*insert
date*] I pointed out that failure to give possession
is a serious matter.

It is something which you must negotiate with the
contractor if you wish to avoid the charge of
repudiation and heavy damages. With your agreement,
I will try to negotiate on your behalf but, since
this is outside the terms of my appointment, I should
be pleased to have your written authorisation to act
for you in this way and your agreement to pay my
additional fees and costs on a time basis as laid
down in my original conditions of appointment.

Yours faithfully

(see section 9.2.2). In practice, you should issue the certificate immediately.

Note that it is your opinion which is required by the contract, not that of the employer or the contractor. Despite the significant consequences of your certificate, 'practical completion' is nowhere defined in the contract. It does not mean substantially or almost complete and the precise meaning has exercised several judicial minds. On balance, it seems to mean the stage at which there are no defects apparent and only very trifling items remain outstanding (*W. Nevill (Sunblest) Ltd* v *Wm Press & Son Ltd* (1981) 20 BLR 78). It is not thought that you are justified in withholding your certificate until every last screw and spot of paint is in place. That would indeed be completion, but the contract clearly intends something rather short of that by the use of the word 'practical'. Within these guidelines you are free to exercise your discretion. You would not be justified in issuing your certificate, in any event, if items remained to be finished which would seriously inconvenience the employer.

You are under no obligation to issue lists of outstanding items if you withhold your certificate. Clerks of works often consider it part of their duty to supply the contractor with so called 'snagging lists'. It is a bad practice because the contractor tends to assume that when he has completed the lists, you will issue a practical completion certificate and disputes sometimes occur. The contractor's obligations should be clear from the contract documents and the duty to make sure that the work is complete in accordance with the contract lies with the contractor not with you or the clerk of works (if employed). More particularly, any 'snagging lists' should be prepared by the contractor's person-in-charge as part of his normal supervision of the works. Whether he does or not, it is not your concern.

Your duty to issue a practical completion certificate does not depend upon any request by the contractor. You must carry out your duty as soon as you are satisfied that the works are in the required state. Many architects arrange a handover meeting to which the employer and any consultants are invited. Remember, however, that you cannot transfer responsibility for certifying practical completion to the employer although it may be a prudent move to see that he is happy with the building before he takes possession. In practice, it is much more likely that the employer will wish to take possession before you are thoroughly satisfied. In such circumstances you must strictly observe your duty and refuse to issue your certificate until you are so satisfied. The contractor will then gain no benefit and may be at some disadvantage in completing the work. He may complain to the employer who may, in turn, complain to you. Put your position to him in writing for the record (Fig 9.3). If you submit to pressure, you may well leave yourself open to a future claim for negligence.

Fig 9.3
Letter from architect to employer: if he has taken possession of the works before practical completion


```
PROJECT TITLE

Dear Sir

I refer to your letter of the [insert date].

I confirm that, in my opinion, practical completion
has not been achieved.  Therefore, it is my duty
about which I have no discretion, to withhold my
certificate.  I note, however, that you have agreed
with the contractor to take possession of the
building.  Although I think you are unwise, I respect
your decision and I will continue to inspect until I
feel able to issue my certificate.  At that date, the
defects liability period will commence.  Naturally,
the contractor has a very great interest in obtaining
a certificate of practical completion and you must
expect him to continue to complain until, in my
opinion, practical completion is achieved.


Yours faithfully
```

9.2.2 Consequences of practical completion

The issue of your certificate of practical completion is of particular importance to the contractor because it marks the date when:

○ The defects liability period commences (clause 2.5).

○ The contractor's liability for frost damage ends (clause 2.5).

○ The contractor's liability for liquidated damages ends (clause 2.3).

○ The employer's right to deduct full retention ends and half the retention held becomes due for release within 14 days (clause 4.3).

○ The machinery culminating in the issue of the final certificate is set in motion (clause 4.4).

○ The contractor's liability to insure under clause 6.3A ends.

9.3 Defects liability period

9.3.1 Definition

The defects liability period is inserted for the benefit of both parties. It allows a period of time for defects to appear and be corrected with the minimum of fuss. Any defect is a breach of contract on the part of the contractor and, without such a period, the employer would have no contractual remedy. He would be left to his common law rights. Moreover, if there were no defects liability period, the contractor would have no right to re-enter the site to remedy the defects. If the employer suffers some loss as a direct result of the defects and the mere remedying of those defects is not adequate restitution, it is always open to him to take action at common law to obtain damages from the contractor for the breach.

Contractors commonly hold two mistaken views about the defects liability period:

○ That the contractor's liability for remedying defects ends at the end of the defects liability period.

○ That the contractor is liable to correct anything which is showing signs of distress or with which the employer is not satisfied.

The contractor's liability does not end at the end of the defects liability period. What does end is his privilege to return and remedy his breach. After that time, the contractor remains liable, but the employer can simply pursue an action at common law for damages, if he so wishes, without giving the contractor the opportunity to return.

The second mistaken view probably owes its origin to the practice of refering to the defects liability period as the 'maintenance period'. Architects and contractors alike are guilty in this respect (even GC/Works/1 and ACA 2 use the term) although maintenance implies a heavier responsibility than simply making good defects. Repolishing, cleaning and gen-

erally keeping a building in pristine condition could be said to be maintenance for which the contractor has no responsibility. The employer's dissatisfaction with the building is also of little consequence in itself (see
section 9.3.2) if the contractor has carried out his obligations.

9.3.2 Defects, excessive shrinkages and other faults

Clause 2.5 requires the contractor to make good 'defects, excessive shrinkages and other faults'. Case law has established that the phrase 'other
faults' must be interpreted *ejusdem generis* with defects and excessive shrinkages; that is to say, faults of the same kind. Read in this way 'other faults'
appears to add little if anything to the contractor's liability. A defect can
only mean something which is not in accordance with the contract. If the
employer is unhappy about the paintwork, it could be that the contractor
has not applied it correctly in accordance with the specifications. On the
other hand, it could be that the specification is inadequate. Only the former situation would give rise to liability on the part of the contractor. An
inadequate specification is usually your responsibility unless it is so inadequate that the finished product would not be reasonably fit for its intended
purpose, when the contractor would have some liability in the matter (see
Chapter 5). The difference between three or four coats of paint, however,
would rarely fall into that category.

It is clear that not all instances of shrinkage fall within the defects liability
clause. Shrinkages have to be excessive. The intention appears to be to
exclude those shrinkages which could be said to be an unavoidable consequence of building operations. This is an eminently sensible approach,
but one which could result in dispute because what is excessive to you may
be trifling to the contractor. Indeed, the whole question of shrinkages is
fraught with difficulty. They are the contractor's liability only if they
result from workmanship or materials not in accordance with the contract. In practice, since the employer holds the purse-strings, it is the contractor who has to convince you that the shrinkage is not his responsibility. Shrinkage usually occurs due to loss of moisture or thermal
movement. A common example is the shrinkage which occurs in timber
after the building is heated. You will have specified a maximum moisture
content for the timber, and subsequent shrinkage can only be because the
timber was installed with, or allowed to develop, too high a moisture content, or your specification was wrong. It is a question of fact rather than
law, but note that it is no defence for the contractor to say that your
specified moisture content was impossible to achieve under normal site
conditions. It is well known that it is difficult to maintain a low moisture
content under the normally damp conditions which prevail on site, but
that is not to say that such conditions cannot be improved by the use of
suitable temporary heaters, ventilation, etc. The contractor's obligation is

to provide workmanship and materials in accordance with the contract and he would have been well aware of your requirements at tender stage. What he is really saying, therefore, is that he found it too expensive to comply with your specification, and that is no defence at all.

9.3.3 Frost

The contractor's liability to make good frost damage is limited to damage caused by frost which occurred before practical completion. This is perfectly reasonable since he was in control of the works up to, but not after, practical completion. Damage due to frost occurring after practical completion is the responsibility of the employer. In practice, there should be no great difficulty in detecting the difference. Frost damage after practical completion may be due, for example, to faulty detailing, unsuitable materials or lack of proper care by the employer. Note that the test is not when the damage occurred, but when the frost, which resulted in such damage, occurred.

9.3.4 Procedure

The defects liability period starts on the date of practical completion as stated in your certificate. If you do not insert any period of time in the contract, the period will be three months. There is really no good reason for limiting the period to three months because there is no connection between the contract period and the defects liability period. Whether the contract period is long or short, there will be some items of work completed just before practical completion and it is these items which you must consider when advising the employer of the length of defects liability period required in a particular case. In general, there is probably much to be said for inserting twelve months as the length of the period on the basis that the building will be tested against all four seasons of the year. It is probably true that the contractor will include a slightly increased tender figure if you include twelve months as the defects liability period, but there is really no reason why he should do so. It probably stems from a mistaken idea of the limits of his liability (see section 9.3.1). The final certificate will, of course, be delayed, but only 2½% of the retention will be outstanding and payment will then probably coincide with your certificate that all defects have been made good (see Chapter 11).

The defects, etc., which the contractor is to make good are those 'which appear within' the defects liability period. The wording suggests that any defects which have already appeared before practical completion could not be included as defects which the contractor must make good. In practice, the situation is not as bad as that. Defects which were apparent (sometimes referred to as 'patent defects') before practical completion would preclude you from issuing your certificate of practical completion (see section 9.2.1). Moreover, since no certificate is conclusive under this contract, the contractor's obligation to carry out the work in accordance with the contract is not reduced by the issue of any such certificate and if the contractor is so misguided as to refuse the opportunity of remedying the defects during the defects liability period, the employer retains his common law rights intact. The danger is that if you certify practical completion while there are some patent defects, half the retention will be released and the contractor may never return. The employer's common law rights will be useless if the contractor has gone into liquidation and the employer may, rightly, say that you were negligent in the issue of your certificate and look to you for recompense. If you do overlook some defects before issuing your certificate, the sensible thing to do is to include them with the defects which actually appear 'within' the period. This may not be precisely what the contract says, but it does no violence to the contractor's rights.

Unlike the equivalent clause (17) in JCT 80, no procedure is laid down for you to issue a schedule of defects or even notify the contractor that defects exist. However, since the contractor cannot know of any defects which may appear within the period unless you do notify him, it is thought that you can use your powers under clause 3.5 to instruct the contractor to carry out urgent remedial action during the defects liability period and to issue an instruction at the end of the period listing all the defects which have appeared during the period. It is prudent to organise an inspection a few days before the period expires so that you can issue your final list of defects on the final day.

There is no time limit set for the contractor to make good the defects, but you are not obliged to certify that making good has been achieved until you are satisfied (to do otherwise would be negligent) and the issue of the final certificate (clause 4.4) is dependent upon your certificate under clause 2.5. The contractor must carry out his obligations within a reasonable time. What is reasonable will depend upon:

○   The number and type of defects.

○   Any special arrangements to be made with the employer with regard to access.

If the contractor fails to carry out his obligations, you may put the compliance procedure under clause 3.5 into operation (see section 4.3).

All defects are to be made good by the contractor entirely at his own cost 'unless the Architect shall otherwise instruct'. Some commentators have been greatly concerned by this last phrase, even going so far as to suggest that it can only mean that the architect can instruct the contractor to remedy defects at the employer's expense. With great respect, we suggest that such a singular view is nonsense. There is no doubt that the wording could be clearer, but it covers two distinct situations which might arise:

○   If the defects are partly the fault of the contractor and partly contributed to by some default of you or the employer. In such circumstances it would be wrong to expect the contractor to remedy the defects entirely at his own expense and you might wish to direct him regarding the extent to which he would be expected to bear the cost. The key words in the clause are 'entirely' and 'unless'.

○   If the employer does not wish the contractor to remedy certain defects because, for example, to do so would seriously disrupt the employer's business. In such circumstances, you might wish to instruct the contractor not to make good such defects.

In either of the above situations, you must be sure to discuss the matter thoroughly with the employer before you issue any instruction. In the second case, you must obtain a letter from the employer authorising you to instruct the contractor that making good is not required (Letter 9.4). If you so instruct, note that there is no provision for you to authorise any deduction from the contract sum.

Although clause 2.5 does also empower you (as other commentators suggest) to instruct the contractor to make good defects at the employer's expense, it is not something which you should ever consider doing.

When making good has been completed you must issue a certificate to that effect. This certificate has important implications with regard to the issue of the final certificate.

9.4   Summary

Possession

○   Must be given to the contractor in sufficient time for him to complete the works.

○   Cannot be revoked by the employer under the general law, but the contract may give such power.

○   Is the same as the date on which works may be commenced in MW 80.

○   Is the contractor's right, and failure on the part of the employer is a serious breach of contract.

○   Ends at practical completion.

Practical completion

○   Is not defined and is a matter for your opinion alone.

○   Requires you to issue a certificate, there is no provision for sectional completion or partial possession.

○   Bestows important benefits on the contractor.

Fig 9.4

Letter from architect to employer: if some defects are not
to be made good

```
PROJECT TITLE

Dear Sir

I understand that you do not require the contractor
to make good the following defects:
[List].

These defects are included in my schedule of defects
issued at the end of the defects liability period.
In order that I may issue the appropriate
instructions in accordance with clause 2.5, I should
be pleased if you would confirm the following:

1. You do not require the contractor to carry out
   making good to the defects listed in this letter.

2. You waive any rights you may have against any
   persons in regard to the items listed as defects
   in the above-mentioned schedule of defects and not
   made good.

3. You agree to indemnify me against any claims made
   by third parties in respect of such defects.

Yours faithfully
```

Defects liability period

o   Benefits the contractor.
o   The end of the period does not mean the end of the contractor's liability to remedy defects.
o   Defects are severely limited in scope.
o   May be of any length.
o   Is not a maintenance period.
o   Only those defects appearing within the period are covered under the contract.
o   Defects must be notified to the contractor during or at the end of the period.
o   Defects must be made good at the contractor's own cost.
o   You have power to instruct the contractor not to make good some defects or to instruct what proportion of cost must be borne by the contractor; check with employer first.
o   When making good is complete, you must issue a certificate to that effect.

Chapter 10

Claims

10.1   General

It is traditional in the construction industry for claims by the contractor
for both extra time and extra money to be linked together. That pattern is
followed in this chapter, but you should bear in mind that there is no
necessary link between time and money. There can be money claims for
both prolongation and disruption, and claims give rise to hard feelings.
MW 80 is unique among the JCT contracts in that it contains no clause
entitling the contractor to 'direct loss and/or expense'. But the absence of
such a clause should not mislead you into thinking that the contractor
cannot make financial claims. He can; but such claims must be pursued at
common law by way of arbitration or litigation and must be based upon
breach of some express or implied term of the contract or some other legal
wrong. You cannot deal with such claims, unless the employer expressly
authorises you to do so, and the contractor agrees. However, you may be
the cause of such claims if, for example, you are late in issuing the contractor with necessary information or instructions, or if you fail to carry out
your duties under the contract and the contractor suffers loss as a result.
The recent case of *Croudace Ltd* v *London Borough of Lambeth* (1986) 6 Con
LR 72 establishes this point and the judgment emphasises that in carrying
out your duties under the contract you owe a duty to both employer and
contractor to act fairly, as you do when certifying: *Sutcliffe* v *Thackrah*
[1974] AC 727. If you fail to carry out your duties properly the case points
out that this is a breach of contract for which the employer is liable in
damages. If that happened, no doubt the employer would seek to recover
from you – and you would have to pay.

10.2   Extension of time

10.2.1 Legal principles

At common law, the contractor is bound to complete the work by the date for completion stated in the contract, unless he is prevented from doing so by the employer's fault or breaches of contract – and the employer's liability extends to your wrongful acts or defaults within the scope of your authority. In the absence of an extension of time clause, neither you nor the employer would have any power to extend the contract period.

Clause 2.2 deals with extension of the contract period and is linked with clause 2.3 which provides for liquidated damages. The only effect of your extending the contract period is to relieve the contractor from paying liquidated damages at the stated rate for the period in question.

10.2.2 Liquidated damages

There are many misconceptions about liquidated damages. The most important point is that the sum stated as such is recoverable whether or not the employer can prove that he has suffered any loss as a result of late completion or even if, in the event, he suffers no loss at all.

Although often referred to by contractors as 'the penalty clause', a penalty is invalid. A sum is treated as liquidated damages (and so recoverable) if it is a fixed and agreed sum which is no more than reasonable and is a genuine pre-estimate of the loss likely to be incurred, or a lesser sum, estimated at the time the contract is made.

Under MW 80 the amount inserted as liquidated damages is usually relatively small in relation to the potential loss to the employer from late completion, but it is bad practice to pluck a figure out of the air. You should have calculated the amount of liquidated damages carefully at pre-tender stage.

To set the figure at the right level, you should discuss it carefully with your client and then make a calculation. This is sometimes difficult. In the case of profit-earning assets, there is no problem. All you need to do is to analyse the likely losses and additional costs. The following should be considered:

o   Loss of profit on a new building, e.g., rental income, retail profit, etc.

o   Additional supervision and administrative costs – including additional professional fees.

o   Any other financial results of the late completion e.g., storage charges for furniture in the case of a domestic building.

An alternative method is to use one or other of the several formulae used in the public sector, such as that put forward by the Society of Chief Quantity Surveyors in Local Government, which gives an approximation to a detailed analysis of all individual costs.

The figure thus arrived at must be inserted in clause 2.3. If no figure was inserted, no liquidated damages would be payable and in the event of late completion the employer would be left to sue for unliquidated damages at common law, based on his proven loss. The same result would follow if no completion date was inserted in clause 2.1 since there must be a date from which liquidated damages can run: *Kemp* v *Rose* (1858) 1 Giff 258.

Clause 2.3 makes it clear that liquidated damages are payable at the specified rate only if the works are not completed by the original completion date or extended contract completion date. They are payable by the contractor 'for every week or part of a week during which the Works remain uncompleted'. The clause does not confer an express right to deduct liquidated damages from monies due to the contractor, e.g., progress payments, but it is plain that this right exists under the general law on the principles stated by the House of Lords in *Modern Engineering (Bristol) Ltd* v *Gilbert-Ash (Northern) Ltd* [1974] AC 689.

10.2.3 Extending the contract period

Unlike the long and complicated provisions in other JCT contracts, clause 2.3 is short and sweet, since it empowers you to extend the time for completion 'if it becomes apparent that the Works will not be completed . . . *for reasons beyond the control of the contractor* . . .' It is not clear whether the italicised phrase in fact extends to delay which is the fault of the employer; sensibly it ought to do so, but the case law suggests otherwise.

In *Peak Construction (Liverpool) Ltd* v *McKinney Foundations Ltd* (1970) 1 BLR 111, the Court of Appeal ruled that liquidated damages clauses and extension of time clauses were both to be interpreted strictly *contra proferentem* against the employer and should not be read to mean that an employer can recover damages for delay for which he was partly to blame. It was said that 'if the employer is in any way responsible for the failure to achieve the completion date, he can recover no liquidated damages at all and is left to prove such general damages as he may have suffered'.

That case concerned an in-house form of contract, but its authority has recently been upheld by the Court of Appeal in *The Rapid Building Group Ltd* v *Ealing Family Housing Association Ltd* (1985) 1 Con LR 1, a case which involved a JCT 63 contract.

Flowchart Fig 10.1 illustrates the contractor's duties in claiming an extension of time under clause 2.3.

Flowchart Fig 10.2 sets out your duties in relation to such a claim.

The procedure under clause 2.3 is straightforward:

The contractor must notify you – not necessarily in writing – 'if it becomes

Flowchart 10.1
Contractor's duties in claiming an extension of time

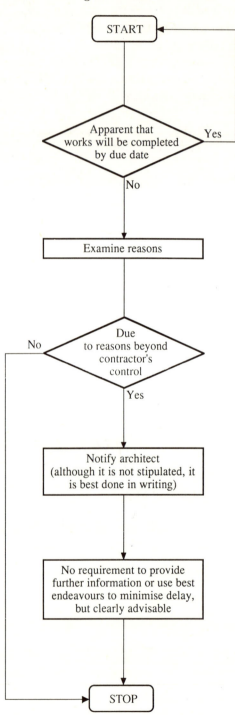

Flowchart 10.2
Architect's duties in relation to claim for extension of time

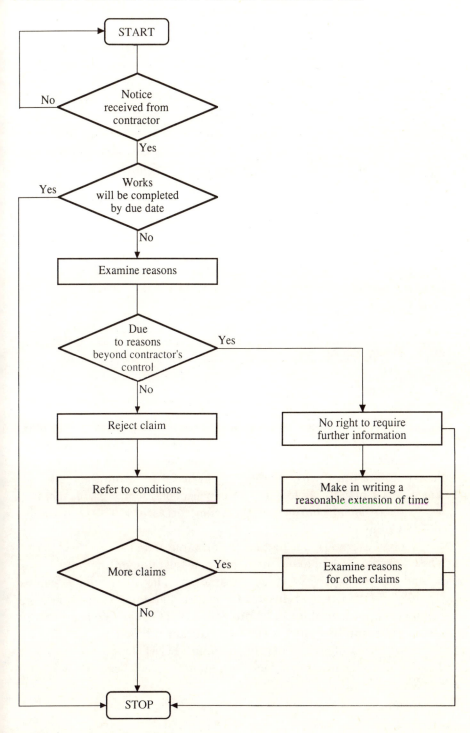

apparent' that the current completion date will not be met. The contract refers to the fact that 'the Works will not be completed by the date for completion'. There is no reference to delays to progress.

You must then make, in writing, a *reasonable* extension of time. The contract does not say when you are to do this, but we suggest that you must grant your extension of time as soon as possible. Certainly you should not spike the application and you must grant the extension before the current date for completion is passed. You must, of course, be satisfied that the completion date will not be met because of 'reasons beyond the control of the contractor'. Failure properly to grant an extension of time may result in the contract completion date becoming at large, so you must be careful. Fig 10.3 is a suggested letter awarding an extension.

What happens if the contractor fails to notify you that the completion date will not be met? Clearly he is in breach of contract in not doing so because the obligation rests on him, but it is unclear whether you have power to extend the contract period if you receive no notification. This is certainly so on a strict reading of the wording of the clause, although common sense might suggest otherwise, and we suggest that in an appropriate case you should award an extension of time even if the contractor has not notified you.

10.3 Money claims

10.3.1 General

There is no clause in the contract which entitles the contractor to make financial claims against the employer whether arising from prolongation or disruption. This leads some people to suppose that this is a risk which the contractor must price. Nothing could be further from the truth.

The absence of a 'direct loss and/or expense' clause merely means:

o   There is no contractual provision for ascertaining and paying money claims.

o   You have no power to quantify or agree claims.

The contractor must pursue his claims in arbitration or litigation, unless he and the employer can agree the amount. The absence of a 'claims' clause is a grave defect of this form of contract and not a benefit as some employers appear to think. The problem can, of course, be overcome by including a suitably worded clause as an amendment to the contract. There are plenty of precedents, for example, IFC 84 clauses 4.11 and 4.12. This again is something which you should have discussed with the employer.

Fig 10.3
Letter from architect to contractor: granting extension of
time under Clause 2.2

```
PROJECT TITLE

Dear Sirs

On [insert date] you notified me that the works would
not be completed by [insert clause 2.1 date or
extended date] because of [specify reasons e.g., the
national strike of building trades employees in
pursuit of a pay claim].

I accept that this falls within clause 2.2 of the
contract, and in accordance therewith I hereby grant
you an extension of time of [specify period]. The
revised date for completion is now [insert date].

Yours faithfully
```

10.3.2 Types of claims

Financial claims are commonly referred to in the industry as *ex-contractual claims*. These are claims made outside the express provisions of the contract, usually as a result of breach of its terms, express or implied. They are also called common law claims, and increasingly are based on *implied terms* relating to co-operation and non-interference by the employer.

The situation is well-illustrated by *Holland Hannen & Cubitt (Northern) Ltd v Welsh Health Technical Services Organisation* (1981) 18 BLR 80 where it was pleaded on behalf of contractors under a JCT 63 contract that there were implied terms whereby the employer contracted that he and his architect:

o   Would do all things necessary on their part to enable [the contractors] to carry out and complete the works expeditiously, economically and in accordance with the contract.

Conversely, that neither the employer nor his architect:

o   Would in any way hinder or prevent [the contractors] from carrying out and completing the works expeditiously, economically and in accordance with the contract.

We suggest that such terms are to be implied when the contract is in MW 80 form and this being so opens up a wide area for claims, which may also arise in tort.

If claims of this nature are made by the contractor they must be made against the employer. He may well seek your advice, and possibly authorise you to deal with them. But the contractor is not entitled to reimbursement because he is losing money; it is up to him to establish that the employer or the architect is in breach. Valid claims can only arise because the contractor suffers loss through your fault or the fault of the employer.

Table 10.4 summarises the MW 80 clauses which may give rise to claims. Useful books for reference purposes are:

o   *Contractor's Claims – An Architect's Guide*, by David Chappell, The Architectural Press, 1984

o   *Building Contract Claims*, by Vincent Powell-Smith and John Sims, Collins, 1985

o   *The Presentation and Settlement of Contractors' Claims*, by Geoffrey Trickey, Spon, 1983.

Table 10.4

Clauses that may give rise to claims under MW 80

Clause	Event	Type
1.2	Failure to issue further necessary information	CL
2.1	Failure to allow commencement on the due date by lack of possession or otherwise	CL
2.2	Failure to give extension adequately or in good time	CL
2.3	Wrongful deduction of damages	CL
2.4	Failure to certify practical completion at the proper time	CL
2.5	Wrongful inclusion of work not being defects, etc. Failure to issue certificate that the contractor has discharged his obligations	CL CL
3.1	Assignment without consent	CL
3.2	Unreasonably withholding consent to sub-letting	CL
3.4	Unreasonably or vexatiously instructing removal of employees from works	CL
3.5	Failure to confirm oral instructions Instructions altering the whole character or scope of the work Wrongful employment of others to do work	CL CL CL
3.6	Variations	C
4.1	Errors or inconsistencies in the contract documents	C
4.2	Failure to certify payment if requested	CL
4.3	Failure to certify payment at practical completion	CL
4.4	Failure to issue final certificate	CL
4.5	Contribution, levy and tax changes	C
5.1	Divergence between statutory requirements and contract documents or architect's instruction	C
7.1	Invalid determination Payment after determination	CL C
7.2	Payment after determination	C

KEY
C = Contractual claims   CL = Common law claims
Contractual claims are usually dealt with by the architect.
Common law claims are usually dealt with by the employer.
NOTE: There is no loss and/or expense clause in this contract. Such claims have to be made at common law.

10.4 Summary

Claims for time and money are distinct; there is no necessary connection between the two.

Liquidated damages

Liquidated damages are:
o A genuine pre-estimate of likely loss or a lesser sum.
o Recoverable without proof of loss.
o Recoverable only if the contractor has not completed the works by the original or extended completion date.

Extension of time

You are bound to grant a reasonable extension of time for completion if:
o The contractor notifies you that the works will not be completed by the current completion date because of reasons beyond his control.
If you fail to grant an extension of time for completion, the contract time may become 'at large' and liquidated damages will be irrecoverable.

Money claims

o There is no contractual provision for the contractor to be reimbursed 'direct loss and/or expense'.
o Any claims by the contractor must be pursued under the general law.
o Such claims can only arise because the contractor suffers loss through your fault or that of the employer.
o Consider inserting a suitable claims clause.

Chapter 11

Payment

11.1 Contract sum

The contract sum is the sum of money which is inserted in Article 2 of the Articles of Agreement. It is stated to be exclusive of VAT which means that VAT payments which may be necessary will be additional to this sum (clause 5.2). How far this will affect the employer will depend upon his status, from the point of view of being able to reclaim VAT, and the work involved in the contract. The employer's attention should be drawn to his liability for VAT; it is something which tends to be forgotten when tenders are received.

The figure in Article 2 will be the contractor's tender figure or such figure as the parties agree, perhaps after negotiation. The importance of the sum cannot be over-emphasised. It is the sum for which the contractor has agreed to carry out the whole of the works as shown upon the contract documents. MW 80 is a 'lump sum contract' which means that the contractor is entitled to payment provided he completes substantially the whole of the works. In theory, the existence of a system of interim payments does not alter the position and if the contractor abandons the work before completion, the employer is entitled to pay nothing more.

Once written into the contract, the contract sum may be adjusted only in accordance with the contract provisions (Table 11.1). Errors or omissions in the computation of the contract sum are deemed to be accepted by employer and contractor. Inconsistencies may be corrected in accordance with clause 4.1 (see section 3.1.2). It is quite possible for the contractor to make a considerable error in his calculations; to such an extent that the contract becomes no longer viable from his point of view. If the error is undetected before the contract is entered into, there is nothing the contractor can do about the situation, except perhaps submit an *ex-gratia* (on grounds of hardship) claim, with little hope of success. You must try to avoid the situation because it is unsatisfactory from all points of view. The

Table 11.1
Adjustment of contract sum under MW 80

Clause	Adjustment
2.5	Defects, etc., during the defects liability period
3.6	Variations
3.7	Provisional sums
4.1	Inconsistencies in the contract documents
4.4	Computation of the final amount
4.5	Contribution, levy and tax changes
6.3A	Insurance money
6.3B	Making good of loss or damage

employer may indeed think that he is gaining, but a contractor in this position has very little incentive to work efficiently and every reason to submit claims at every opportunity. Unfortunately, unless your contract documentation includes quantified schedules or bills of quantities, it is very difficult to check the contractor's pricing. Some errors may be obvious, but where the contract documents consist of drawings and specification, the contractor's pricing strategy may be obscure unless he separately submits a detailed breakdown of the figure.

11.2   Payment before practical completion

Clause 4.2 sets out the procedure for what is termed 'progress payments'. You are to certify 'progress payments to the contractor' at not less than four-weekly intervals. The phrase is ambiguous. It may mean that you are to certify *progress payments to the contractor*, such certificates being sent, as is now customary, to the employer with a copy to the contractor; or it may mean that you are to certify *progress payments* and send the certificates to the contractor. On balance, it is suggested that the latter interpretation is correct since there is no other indication to whom the certificate should be sent. It is essential, however, that you send a copy to the employer by registered post because his liability to pay is not stated to be conditional upon presentation of the certificate by the contractor. The employer must pay within 14 days of the date of the certificate.

You are not obliged to issue your certificate unless the contractor specifically requests it. In practice, it is usual for the contractor to send four-weekly statements of account for your consideration.

Although there is no provision for any other system of payment, it may be more convenient, on small works, to agree a system of stage payments. If it is desired to operate in this way, it is important to make the necessary amendments to the printed conditions and to ensure that the contractor is aware of the change at tender stage. It will have a considerable impact on his pricing strategy.

The amount to be included in your certificate is to consist of:

○   The value of works properly executed.

○   Amounts ascertained or agreed under clause 3.6 (variations) or clause 3.7 (provisional sums).

○   The value of any materials and goods which have been reasonably and properly brought upon the site for the purpose of the works and which are adequately stored and protected against the weather and other casualties.

Less:

○   A retention of 5%.

○   Any previous payments made by the employer.

Each of the above items requires careful consideration:

The value of works properly executed

There has been a difference of opinion about the meaning of the word 'value' in this context. The generally accepted view is that it is the valuation obtained by means of reference to the priced document in relation to the amount of work actually carried out. Defective work is not included and a retention percentage is deducted. The retention is intended to deal with problems which might arise. The biggest problem would be the contractor's going into liquidation immediately following a payment. The alternative view of the meaning of 'value' derives from this possibility. From the employer's point of view, the value of the contractor's work is the value of the whole contract less the cost of completing the work with the aid of another contractor and additional professional fees. The additional cost would be considerable and incapable of being met from the retention fund.

The latter view finds little favour with contractors since the certificates issued at any given stage would bear no relation to the money expended by the contractor. Indeed, during the early part of a job, the certificates might even show a minus figure. The two views might be termed contractor value and employer value. The services of a quantity surveyor would be necessary to determine the probable cost of completion at the time of each certification. His task would not be easy. It is possible to operate this system if you make it clear to the contractor at the time of tender so that he can take extra financing charges into account. Recent case law suggests that the courts will favour the traditional contractor value view of 'value' if you do not state in express terms that an alternative interpretation is to be used.

'Properly executed' refers to the fact that you are not to include the value of work which is defective, that is, not in accordance with the contract. If you do certify defective work because, perhaps, the defect does not make itself immediately apparent, the correct procedure is to omit the value from the next certificate. This should pose no difficulties unless, of course, the contractor abandons the work first.

Amounts ascertained or agreed under clauses 3.6 and 3.7
This refers to any variations which are valued before the date of the certificate and any instructions which you may issue with regard to provisional sums which result in an adjustment to be valued under clause 3.6.

The value of materials and goods, etc.
The reasoning behind the inclusion of payments for unfixed materials on site is clear. It is to enable the contractor to recover, at the earliest possible time, money which he has already laid out. You need not include any materials which you consider have been delivered to the site unreasonably early for the sole purpose of obtaining payment, but in its present form the clause contains serious dangers for the employer. There is no provision for the contractor to provide proof of ownership before payment. Therefore, if unfixed materials are included in your certificate and the contractor does not own them, the employer could be faced with the prospect of paying twice for the same materials if the contractor goes into liquidation and the true owner claims the goods from site (*Dawber Williamson Roofing Ltd* v *Humberside County Council* (1979) 14 BLR 70). The prevalence of retention of title clauses in the supply contracts of builders' merchants makes this a very real danger. Such a retention of title clause cannot be overcome by anything which you may write into the contract. It is important that you amend the clause, preferably by deleting the whole of the provision 'and the value of any materials or goods . . . protected against the weather and other casualties'. Since, fortunately, there is no provision for payment for off-site materials, this will then mean that you do not have to certify any unfixed materials whatsoever. It should pose no problems in practice provided that you make the point clear to the contractor at the time of tender. The slight increase which the contractor will make to his tender figure is well worth it and can be considered in the same light as an insurance premium.

A retention of 5%
This is covered in detail in section 11.6.

Any previous payments made by the employer
The difficulty here is that, at the time of issuing a certificate, you have no automatic way of knowing whether the employer has made previous

payments. This clause puts the onus on you to find out. There is the added difficulty that the employer may have legitimately set-off sums against previous certificates and your calculations could become very complicated. As it stands, the clause binds you to make enquiry of the employer and the contractor before the issue of each certificate. If they disagree, you are in an awkward position. The simple way out of the problem is to amend the phrase by omitting 'any previous payments made by the employer' and adding 'amounts previously certified'. This brings the provision into line with JCT 80 and IFC 84.

11.3 Penultimate certificate

The contract provides for a special payment to be made at practical completion (clause 4.3). This clause is titled 'penultimate certificate' because, although it is not expressly stated, it is clear that regular certificates will cease at this point because there is no more work to value. Indeed the only other payment to be made will be the final payment.

You are to issue the penultimate certificate within 14 days of the date you certify practical completion. The employer has 14 days, as before, in which to pay. The effect of the certificate is to release to the contractor the balance of the monies due on the full value of the works, plus half the retention already held. The employer retains only 2½% of the total value. Obviously, it may not be possible for you to know the full value precisely at this stage, but you are to certify it 'so far as that amount is ascertainable at the date of practical completion'. In other words, you are to do the best that you can pending final computation. Amounts ascertained or agreed under clauses 3.6 and 3.7 are to be included and, again, reference is made to deducting the amounts of previous progress *payments* made. The comments made earlier are also applicable to this clause.

11.4 Final certificate

The contract lays down a precise time sequence for the events leading up to, and the issue of, the final certificate (see the contract time chart, Fig 11.2).

The contractor's duty is to send you all the documentation you reasonably require in order to compute the amount to be finally certified. You are entitled to request any particular supporting evidence you require. The contractor has three months, or such other period as you insert in the contract, from the date of practical completion to send you the information. You must issue your final certificate within 28 days of the receipt of the contractor's information. There is a proviso. You may not issue your final certificate until after you have issued your certificate of making good defects under clause 2.5. In practice, this will often be the deciding factor. If the contractor is late in sending you the information, he is technically in

Fig 11.2
MW 80 Time Chart

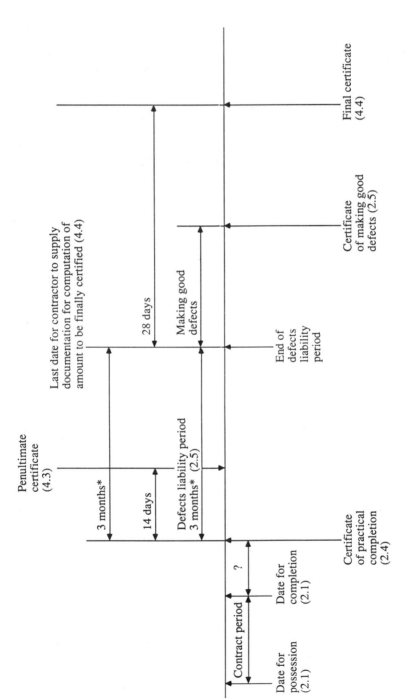

breach, but it is of little consequence. He is simply delaying the time when he receives payment because your 28 days does not begin to run until you receive his documentation.

Your certificate is to state the amount remaining due to the contractor or, more unusually, the amount due to the employer. The latter situation will only occur if you have overcertified – a situation to be avoided. The employer has 14 days, as before, in which to pay from the date of the final certificate.

There is no provision in the contract for the contractor to agree your computations before you issue your final certificate, but it is customary to attempt to obtain his agreement. If the contractor delays sending you his agreement, it does not affect your duty to issue the final certificate within 28 days which remains. Your letter to the contractor should make this clear (Fig 11.3).

11.5   Effect of certificates

It is refreshing to note that no certificate is stated to be conclusive. Thus, the issue of the certificate of making good defects does not preclude you from requiring the contractor to remedy anything which is not in accordance with the contract. Similarly, if you include defective work in a four-weekly progress certificate, you can remedy the situation with the next certificate you issue.

The contractor's liability is not reduced in any way by the issue of the final certificate which is not even conclusive as far as the computations are concerned.

11.6   Retention

The reference to retention in this contract is remarkably brief. It is mentioned only in clauses 4.2 and 4.3 and then only in respect of the amount. In particular there is no reference to:

o   The purposes for which the retention can be used.
o   Its status as trust money.
o   Keeping it in a separate bank account.

The reason for this is clearly to maintain the brevity of the contract as a whole, but there may be repercussions:

Use of retention
Clause 3.5 allows the employer to deduct the cost of employing others to carry out instructions, with which the contractor has failed to comply, from monies due or to become due to the contractor. Retention monies clearly fall within this clause, but the employer will, more commonly, deduct such monies from future certificates. If the contractor goes into liquidation before the work is complete, there is no doubt that the

Fig 11.3

Letter from architect to contractor: requesting agreement
to the computation of the final sum

```
PROJECT TITLE

Dear Sir

I enclose copies of the computation of the final sum
to be certified.

I should be pleased if you would signify your
agreement to the sum and the way in which it has been
calculated by signing and dating one copy of the
calculation in the space provided and returning it to
me by the [insert date] at the latest.

If you have any queries, please telephone me as soon
as possible, but you should note that, in any event,
I have a duty under clause 4.4 of the Conditions of
Contract to issue my final certificate no later than
[insert date].

Yours faithfully


Copy: Employer
```

employer would be entitled to make use of retention monies to complete the work. There is no express provision enabling the employer to make use of the retention for any other purpose.

For example, the employer may not deduct the cost of taking out insurance if the contractor defaults. In practice, as mentioned earlier, the situation may suggest taking a broader view.

Trust money

Unlike the corresponding provisions of JCT 80 and IFC 84, the retention is not expressly stated to be trust money and it is not considered that any term would be implied from the general law to create such a trust.

Whether the retention is or is not a trust is of vital importance to the contractor for two reasons. First, a trust is governed by statute and, despite what the contract provisions may state, it is probable that the employer always has a duty to invest and to return an interest to the contractor. Where the retention is not stated to be a trust, the contractor is not entitled to any interest.

Second, where retention is stated to be trust money, the employer is in the position of trustee; he is merely holding the money for the contractor's ultimate benefit. It is, in no sense, the employer's money. Therefore, if the employer were to become insolvent, the trust money would not form part of his assets in the hands of his trustee in bankruptcy or liquidator. The contractor would be able to recover the full amount. If the money is not a trust, as in the case of MW 80, the contractor would have no better chance of recovery than any other unsecured creditor. MW 80, therefore, leaves the contractor at a severe disadvantage.

Separate bank account

Where the retention is trust money, it is often required to be placed by the employer in a separate bank account specially opened for the purpose. The result is that, in the event of the employer becoming insolvent, there is no difficulty in identifying the money. It is also good practice for any trust money to be dealt with separately from the employer's own money so that the investment income can be easily ascertained. Even if the contract is silent, it is now established law that the contractor can demand that trust money be kept in a separate bank account. In the present case, since the retention is not stated to be trust money, the contractor cannot so demand.

11.7 Variations

Variations are covered by the extremely short clause 3.6. It gives you the power to order:
o An addition to the works.
o An omission from the works.

o   A change in the works.
o   A change in the order in which they are to be carried out.
o   A change in the period in which they are to be carried out.

It should be noted that the contractor has no right to object to any change in sequence or timing of the works.

This seems rather harsh, but taken in the context of the size and type of contract likely to be carried out under this form it is unusual to find problems arising in practice. In any case, the contractor is entitled to be paid for changing the sequence or timing, so he should not, in theory, suffer.

The clause provides that you are to value all the types of variation listed. If a quantity surveyor is employed, you will be wise to let him do this for you, but the contract is silent about the role of the quantity surveyor (see section 7.1) and he has no powers. Even if you do ask the quantity surveyor to undertake the valuation of variations, remember that you bear the ultimate responsibility. So check all valuations carefully.

The contract allows you to agree a price with the contractor for any variation, but agreement must be reached before the contractor carries out your instruction. In practice, it allows you to invite and accept the contractor's quotation. On small works this is the sensible way of operating the provision if the additional or omitted work is anything other than more, or less, of the same.

If you do not agree a price with the contractor, the clause 3.6 lays down how the valuation is to be carried out. It is to be done on a 'fair and reasonable basis'. Although it is pleasant to think that 'fair and reasonable' is an objective concept, in practice it is your opinion as to what is fair and reasonable that matters. When carrying out the valuation, you are to use 'where relevant' prices in the priced documents. They are:

o   The priced specification *or*
o   The priced schedules *or*
o   The contractor's own schedule of rates.

It is a matter for you to decide whether the prices are relevant. It is considered that unless the work being valued is precisely the same, carried out under the same conditions, you are at liberty to ignore the prices in the priced document. This is because even a slight change in the conditions under which work is being carried out will have a marked effect upon the cost to the contractor.

There is no express requirement for the contractor to submit vouchers or other information to assist you in arriving at a fair and reasonable valuation, but it would be a foolish contractor who refused your reasonable requests in this respect. If the contractor refuses to supply information, you must proceed to carry out the valuation to the best of your ability and the contractor will have no valid claim if he receives less than he expects. Your valuation will, of course, include an element for profit, overheads and so on, as usual.

Although not expressly stated, it is suggested that a fair and reasonable valuation would also include the valuation of work or conditions not expressly covered in your instruction but affected by your instruction. For example, if you issue an instruction to change the doors from painted ply-wood faced to natural hardwood veneered and varnished, it might well affect the sequence in which the contractor hangs the doors, the degree of protection required and the difficulty of painting surrounding woodwork. All this must be taken into account in your valuation.

There is no provision in this contract for you to value or ascertain any loss and/or expense caused by any disruption or prolongation of the works. The contractor must claim at common law. In practice, if disruption does occur due, for example, to late delivery of information, the employer may well ask you to try to agree a figure in settlement to avoid court proceedings. The contract, however, makes no provision for so doing.

11.8   Provisional sums

Clause 3.7 empowers you to issue instructions to the contractor directing him how provisional sums are to be expended. Provisional sums are usually included when the precise cost or extent of work is not known at the time of tender. The purpose is to have a sum of money to cover the cost. When you issue your instruction, you must omit the sum and value your instruction in accordance with the principles in clause 3.6 (explained in section 11.7 above). It is possible to use clause 3.7 to nominate a sub-contractor, but you do so at your peril (see section 8.2.3).

11.9   Fluctuations

Clauses 4.5 and 4.6 deal with fluctuations. Clause 4.5 is used if it is intended that the bare minimum fluctuations to deal with contribution, levy and tax changes are to be allowed. Detailed provisions are to be found in Part A of the Supplementary Memorandum. In the case of short-term projects for which this form is intended to be used, this clause will be deleted, as the footnote makes clear.

Clause 4.6 states that, save for the limited provision of clause 4.5 if retained, the contract is to be considered fixed price. The contractor must carry the risk of increases in cost of labour, materials, plant and other resources. Of course, in the unlikely event of a general fall in costs, the contractor would gain.

11.10 Summary

Contract sum

o   Exclusive of VAT.
o   The amount for which the contractor agrees to carry out the whole of the work.
o   MW 80 is a 'lump sum' contract.
o   May be adjusted only in accordance with the contract provisions.
o   Errors are deemed to be accepted by both parties.

Payment before practical completion

o   You are to certify payments at four-weekly intervals *if* the contractor so requests.
o   An alternative system of payment may be agreed between the parties.
o   The employer must pay within 14 days of the date of the certificate.
o   You must decide the meaning of 'value' and inform the contractor at tender stage.
o   The employer may reserve his retention on the whole of the certified sum.
o   There are dangers in certifying unfixed materials.
o   If the contract is not amended, you must ascertain the amounts of previous payments by the employer.

Penultimate certificate

o   Certificate must be issued within 14 days of the date of practical completion.
o   Half the retention must be released.

Final certificate

o   The contractor must send all the documents you require to compute the final sum.
o   The contractor has three months from the date of practical completion to send them.
o   The final certificate must be issued within 28 days of the receipt of contractor's information provided that you have issued your certificate of making good defects.
o   The contractor does not have to agree your computations.

Effect of certificates

○ No certificate is conclusive in any respect.
○ The contractor's liability is not reduced by the final certificate.

Retention

○ The employer may deduct money from the retention fund to pay the cost of employing others in accordance with clause 3.5.
○ The retention is not trust money and need not be kept in a separate bank account.
○ The contractor has no better claim on the retention money than any other unsecured creditor if the employer becomes insolvent.

Variations

○ The contractor cannot object to a change in the sequence or timing of the works.
○ You must check all valuations.
○ You may agree a price with the contractor before he carries out your instruction.
○ Otherwise valuations must be on a fair and reasonable basis using prices in the priced document if relevant.
○ There is no provision for the valuation of claims for loss and/or expense.
○ Provisional sum work must be valued in accordance with clause 3.6.

Chapter 12

Determination

12.1   General

Under the general law, a serious failure by one of the parties to a contract may entitle the other (innocent) party to treat the contract as at an end. But it is not every breach of contract which entitles the innocent party to behave in this way. The breach must be 'repudiatory', i.e., it must be conduct which makes it plain that the innocent party will not perform his obligations, or else it must consist of misperformance which goes to the root of the contract.

Until all the facts have been investigated – and this must take place after the event – it is very difficult to say at what point a breach has occurred which is sufficiently serious to entitle the innocent party to terminate the contract.

In common with most standard building contracts, MW 80 provides for either party to determine the contractor's employment under the contract by going through a prescribed procedure, and in fact the determination clause (7.0) attempts to improve on the common law rights of the parties and, indeed, the clause goes on to specify what the rights of the parties are after there has been a valid determination.

Determination of the contractor's employment is a serious step, and the consequences of a wrongful determination are serious. It is best avoided if possible, and the process is fraught with pitfalls for the unwary. Quite apart from the possibility of an action for breach of contract by one party or the other if things go wrong, the employer is always placed in an impossible position as far as getting the project completed is concerned.

Even if the employer is successful in recovering his costs from the contractor, he can never recover the time which he has lost. The formalities laid down in the determination provisions must be followed exactly if a costly dispute is to be avoided.

You bear a heavy burden since you will have to advise the employer about his rights and the procedure to be followed, and the determination clauses are only too easy to misunderstand. Either the employer or the contractor may exercise the right to determine employment and MW 80 deals separately with determination by the employer and by the contractor.

12.2 Determination by the employer

12.2.1 Grounds and procedure

The procedure for determination is set out in the flowchart (Fig 12.1) while the grounds which may give rise to determination and the procedure to be followed are specified in clause 7.1.

There are three separate grounds for determination in that clause. They are that the contractor:

o   Fails to proceed diligently with the works without reasonable cause (clause 7.1.1); or,

o   Wholly suspends the carrying out of the works, before completion, without reasonable cause (clause 7.1.1); or,

o   Becomes insolvent in one of the ways specified in clause 7.1.2.

Each of these grounds is described as a 'default' although insolvency is not a breach of contract at common law. It is a misfortune.

If the employer, on your advice, decides to determine the contractor's employment, you must ensure that the procedure is followed precisely. This form of contract contains no provision for any preliminary notice before the right of determination is exercised, but we recommend that save in the most exceptional cases the employer should warn the contractor of his intention to determine unless the default is remedied. Fig 12.2 is the sort of letter which you might draft for the employer to send, and we suggest that it be sent by registered post or recorded delivery. This gives the contractor advance warning that the employer may exercise his right to determine. In many cases, that will be sufficient to stop the default immediately.

Before you advise the employer to determine the contractor's employment by notice, you should consider the grounds carefully:

Fails to proceed diligently with the works

This is a breach of the contractor's basic obligation in clause 1.1 which requires him 'with due diligence and in a good and workmanlike manner' to carry out and complete the works in accordance with the contract documents. Whether or not the contractor is proceeding diligently is a factual question. Diligence suggests that the contractor must work carefully and with perseverence and the only guidance from decided cases is largely of a negative nature. If there is a progress schedule, this may be helpful, but it is not decisive, and many people believe that if the contractor is making

Flowchart 12.1
Determination by employer

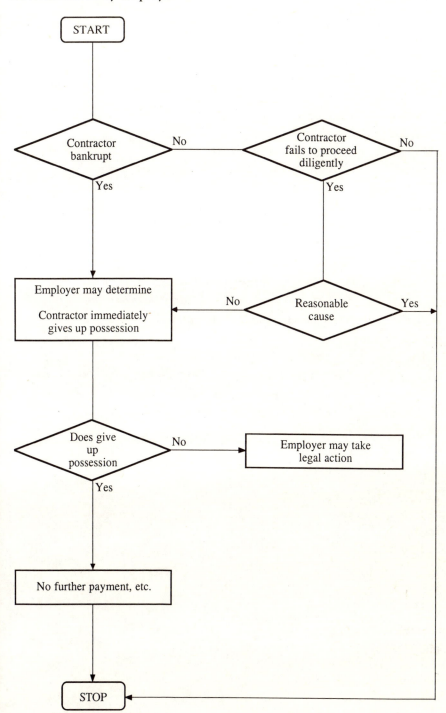

Fig 12.2

Letter from architect to contractor: warning of impending determination

```
REGISTERED POST OR RECORDED DELIVERY


PROJECT TITLE


Dear Sir(s)


When I visited the site today I found that [e.g.,
none of your employees were present and most of your
plant had been removed].  I have spoken to the
employer and drawn his attention to the provisions of
clause 7.1.1. of the above contract.

The employer has instructed me to inform you that
unless you return to site immediately and resume
work, it is his intention to determine your
employment in accordance with clause 7.1  of the
contract.

I trust that this will not be necessary and I will be
glad if you will telephone me immediately on receipt
of this letter stating your intentions.


Yours faithfully



Copy: Employer
```

some progress – however slight – it is dangerous to rely on this ground. Factors which you should take into account include the number of men on site and the number of men required; the amount of plant and equipment in use; what work remains to be done, and the time available for completion; the progress being made and any factors outside the contractor's control which hinder progress.

Wholly suspends the carrying out of the works before completion
'Wholly' means what is says – completely, totally or entirely. The contractor must have completely ceased work, which probably means that he will have left the site and has, in effect, abandoned the work. You should note that his action must be without reasonable cause – and a reasonable cause might well be, for example, the employer's failure to make progress payments under your clause 4.2 certificates.

Financial failure of contractor
Unlike some other contract forms, determination on the grounds that the contractor is insolvent is not automatic. It is by notice. The specified grounds are if the contractor:
○   Becomes bankrupt.
○   Makes a composition or arrangement with his creditors.
○   Has a winding-up order made.
○   Passes a resolution for voluntary winding up (except for purposes of reconstructing the company).
○   Has a receiver or manager appointed.
○   Has possession taken by or on behalf of a creditor of any property which is subject to a charge.
These are all factual matters, and in some cases, e.g., where a receiver is appointed, it may be best to advise the employer not to determine the contractor's employment, since the receiver is bound to carry on with the company's contracts.
The procedure for determining the contractor's employment is that the employer – not the architect – serves a notice by registered post or recorded delivery determining the contractor's employment 'forthwith'. The notice operates from the date it is received by the contractor: *J.M. Hill & Sons Ltd* v *London Borough of Camden* (1980) 18 BLR 31.
You should draft a suitable letter for the employer's signature (Fig 12.3). Determination of the contractor's employment by the employer is governed by an important proviso. The right must not be exercised 'unreasonably or vexatiously'. In simple terms, this means that the procedure must not be instituted without sufficient grounds so as to cause annoyance or embarrassment. In *J.M. Hill & Sons Ltd* v *London Borough of Camden* (1980) 18 BLR at p. 49, the Court of Appeal took the view that 'unreasonably' in this context meant 'taking advantage of the other side in

Fig 12.3
Letter from employer to contractor: determining employment

REGISTERED POST OR RECORDED DELIVERY

PROJECT TITLE

Dear Sirs

Despite the architect's letter to you dated [*insert date of architect's warning letter*] you have failed to resume work on site [*if appropriate*].

In accordance with clause 7.1 of the conditions of contract, take this as notice that I hereby determine your employment under this contract without prejudice to any other rights or remedies which I may possess.

The ground on which this notice is served upon you is [*insert specific details of the default, with dates if appropriate*].

You are required to give up possession of the site of the works immediately, and should you fail to do so I shall instruct my solicitors to issue appropriate Court proceedings against you.

Yours faithfully

Copy: Architect

circumstances in which, from a business point of view, it would be totally unfair and almost smacking of sharp practice'.

12.2.2 Consequences of employer determination

The penultimate sentence of clause 7.1 sets out what is to happen should the employer successfully determine the contractor's employment:

o The contractor must give up possession of the site immediately.
His licence to occupy it is at an end, and if he fails to do so, he becomes a trespasser in law. The contractor must remove himself, his men, plant, materials and tools from site.

o The employer is relieved of any obligation to make any further payment to the contractor until after completion of the works.
Unlike JCT 80 or IFC 84, this contract confers no other rights on an employer who exercises his right to determine – which is another good reason for ensuring that the employer does not act hastily. The employer is given no right to direct loss and/or damage which arises from the determination, nor any rights over the contractor's plant and tools. Equally, the employer has no rights over materials or goods delivered to the works but not certified for payment. In all these respects, therefore, MW 80 is gravely defective.

The only hope lies in the proviso, the effect of which is to preserve the employer's common law rights. To succeed in an action for damages at common law – and so recover his losses – the employer would have to show that the contractor had repudiated the contract, i.e., shown by his conduct he no longer meant to be bound by it, in which case the employer could recover any extra cost to which he has been put by way of damages in arbitration or litigation.

Clearly, the default of wholly suspending the carrying out of the works before completion without reasonable cause is capable of being a repudiatory breach, but neither the contractor's financial failure nor his failure to proceed regularly and diligently with the works meets the common law criteria. The position under MW 80, therefore, is that the employer merely has a right to determine the contractor's employment and nothing more: *Thomas Feather & Co (Bradford) Ltd* v *Keighley Corporation* (1953) 53 LGR 30.

12.3   Determination by the contractor

12.3.1 General

The contractor has a right to determine his own employment for specified defaults by the employer and if he is successful in doing so, the results for the employer will be disastrous. Because of this, you must do everything in

Flowchart 12.4
Determination by contractor

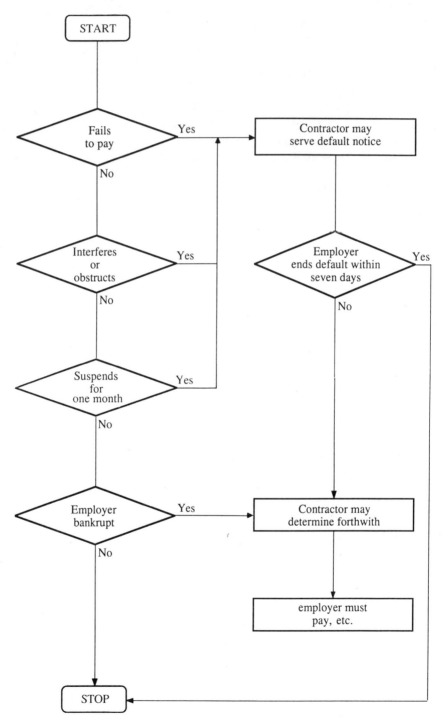

your power to prevent it from happening.

The practical consequences of a successful contractor determination are:

o   The project will have to be completed by others, probably at a higher cost. In any event, professional and other fees will be involved, as well as other expenses.

o   The completion date will be inevitably delayed.

o   The employer may be faced with a common law claim for damages.

The flowchart (Fig 12.4) shows the procedure for determination by the contractor. The grounds and procedure for determination and its consequences are laid down in clause 7.2.

12.3.2 Grounds and procedure

There are five separate grounds which entitlé the contractor to exercise his right to determine his employment under the contract. They are that the employer:

o   Fails to make any progress payment within 14 days.

o   Interferes with or obstructs the carrying out of the works.

o   Fails to make the premises available to the contractor at the right time.

o   Suspends the carrying out of the works for 1 month.

o   Becomes insolvent in one of the ways specified in the clause.

You should note that there is no provision that these defaults must be 'without reasonable cause', and the only safeguard from the employer's point of view is that the contractor must not exercise his option to determine his employment 'unreasonably or vexatiously'.

As in the case of employer determination, the contractor must follow the procedure precisely as it is possible that a wrongful determination would amount to a repudiation of the contract. The procedure differs from that laid down for determination by the employer in that when the contractor wishes to determine on one of the first four of the above grounds, he must first send a notice to the employer (not to you) by registered post or recorded delivery giving notice of his intention to determine. That notice must specify the alleged default. The employer has seven days from receipt of that notice in which to make good the default, and only if he continues the default for seven days may the contractor determine his employment. Determination is brought about by a further notice served on the employer by registered post or recorded delivery and it takes effect on receipt.

In the case of the employer's financial failure (clause 7.2.4) no preliminary notice is required.

It is worthwhile considering the grounds for determination in detail because they have been put in advisedly to protect the contractor against common wrongdoings by employers.

The employer fails to make any progress payments due within 14 days of the due date
This ground protects the contractor's right to be paid on time. Steady cash flow is as important to him as it is to you or your employer and in fact this provision is quite generous to the employer. Under clause 4.2 the employer has to pay to the contractor the amounts which you certify as progress payments within 14 days of the date of your certificate and payment is due to the contractor before that period expires. Under clause 7.2.1 the contractor has to wait a further 14 days – making 28 days from the date you issued your certificate – before he can serve a preliminary notice of intention to determine.

If the employer does receive a notice under clause 7.2.1, he must pay at once, and if you know about it – and you probably will – you should telephone the employer and advise him to pay immediately and confirm this by a letter along the lines of Fig 12.5.

Right from the outset, you must make sure that the employer understands the scheme of payments and the need to pay promptly on your certificates. Where possible, deliver financial certificates by hand and obtain a receipt. If this is impracticable, send them by registered post or recorded delivery. Once the 14 day period of grace has expired, of course, the contractor may sue upon the certificate in any case, although most contractors prefer to rely upon their option to determine employment. If the contractor does issue a writ and applies for summary judgment under Order 14 of the Rules of the Supreme Court he will generally get judgment for immediate payment. The recent case of *C.M. Pillings & Co Ltd* v *Kent Investments Ltd* (1985) 4 Con LR 1 has muddied the waters to some extent because it establishes that if the employer is genuinely disputing the amount of a certificate – and can produce hard evidence to back up his contention – the contractor's action may be stayed and have to go to arbitration. But it is never sufficient for the employer to refuse to pay until he has satisfied himself that it is correct. You bear a heavy responsibility to ensure that it is correct.

The employer or any person for whom he is responsible interferes with or obstructs the carrying out of the works
This is a breach of contract at common law and interference or obstruction is a serious matter. In law, conduct of this kind is often referred to as 'acts of hindrance and prevention' and is a breach of an implied term of the building contract. The reference is to the carrying out of the works. The ground covers a wide range; countermanding your instructions on site, for example, would fall within the clause. You will probably know from the contractor if this sort of thing is happening, and if he does complain to you, and you think that his complaint is justified, you should write to the employer warning him of the likely consequences: Fig 12.6.

Fig 12.5
Letter from architect to employer: advising immediate
payment

PROJECT TITLE

Dear Sir

I refer to our telephone conversation today when I
advised you to make immediate payment to the
contractor of Certificate No.OO amounting to £[*insert
amount*] in light of the service upon you of a notice
preliminary to the contractor determining his
employment under the contract.

You cannot assume that you have a full 14 days to pay
because, technically, the certificate is not honoured
until your cheque has been cleared through the bank.
I suggest that you arrange to have your cheque
delivered by hand to the contractor's office. If you
allow the contractor to determine his employment the
consequences will be considerable extra cost and
delay to the project.

It is essential that you pay the certified progress
payments within 14 days of the date on the
certificate and it is up to you to make the necessary
financial arrangements to ensure that funds are
available at that time.

Yours faithfully

The employer or any person for whom he is responsible fails to make the premises available to the contractor
The use of the word 'premises' is curious as is the wording generally, although the intent is clear. It means the site. Clause 2.1 states quite simply that 'The Works may be commenced on [Date] and shall be completed by [Date]'. There is an implied term in this contract that the employer will give possession of the site to the contractor in sufficient time for him to complete his operations by the contractual date and failure to do so is a serious breach of contract: *Freeman* v *Hensler* (1900) 2 HBC 292.

The employer suspends the carrying out of the works for a continuous period of at least 1 month
This ground is drafted very broadly and it is not the equivalent of, for example, JCT 80, clause 28.1.3, which refers to suspension by *force majeure*, architect's instructions, etc. It is not known why this ground was inserted by the draftsman, because the contract does not expressly confer on the employer any right to suspend the works – and to do so would be a breach of contract – and even if you have the right to order a suspension under your general powers under clause 2.5 it would not fall under this clause. Any suspension must be *continuous* for a period of a calendar month.
Even though the employer has no power to suspend the carrying out of the work he may do so, e.g., because he is in financial difficulties. The contractor could then exercise the right of determination under the clause if that suspension lasted for a month. But he would be better off to rely on his common law rights!

Employer's financial failure
These grounds (clause 7.2.3) parallel those of contractor insolvency in clause 7.1.2 (page 137) and no preliminary notice is required from the contractor. Theoretically, this ground is applicable even in the case of a local authority employer, although the whole clause is drafted on the assumption that the employer is an individual or a limited company since, technically, none of the events referred to is applicable in law to local authorities. This is not to say that local authorities cannot get into financial difficulties or become 'insolvent', as recent events have proved.

12.3.3 Consequences of contractor determination

The first sentence of the last paragraph sets out the contractor's rights to payment following determination of his own employment. The wording is mandatory; the employer must pay to the contractor a fair and reasonable sum for:
o   The value of the work begun and executed.
o   Materials on site.

Fig 12.6

Letter from architect to employer: warning of likely consequences of interference

PROJECT TITLE

Dear Sir

When I visited the site today, the contractor
complained to me that [*insert details as appropriate,
e.g., your office manager insisted that the
contractor's men should not proceed with the
renovation works to the Board Room*].

The contractor intimated to me that he was minded to
determine his employment under clause 7.2 of the
contract on the grounds that this conduct by your
office manager amounts to interference with or
obstruction of the carrying out of the works. He has
previously complained in similar terms.

If the contractor does determine his employment, the
consequences will be serious in terms of both time
and cost. Determination is something to be avoided
at all costs, and I suggest that you advise your
staff that any instructions of this nature could have
serious repercussions for you.

Yours faithfully

o   The cost of removal of all temporary buildings, plant, tools and equipment from site.

This 'fair and reasonable sum' is to be paid over 'after taking into account amounts previously paid'. You are not required to issue a certificate, but we think it is implied that you must endeavour to agree the figures with the contractor on the employer's behalf. If the employer is insolvent, of course, any sums due to the contractor will have to be claimed against the insolvent estate, but if the contractor has determined his employment on other grounds, then his remedy will be to sue. In the absence of an agreed sum, the contractor would be able to bring a *quantum meruit* claim, that is, a non-contractual claim for a 'reasonable sum'.

The contractor does not get any direct loss and/or damage and this means that, under the clause, he cannot recover the profit which he would have made had the job progressed to successful completion. But he may well be able to establish a claim to loss of profit and other consequential losses where the employer's default was a repudiatory breach, and he could then recover those sums in arbitration or litigation.

Effect of determination on other rights

The contractor's right to determine his employment is expressly stated to be without prejudice to any other rights or remedies which he may possess. This means that his ordinary rights at common law are preserved, as are those of the employer. This makes it plain that, for example, the contractor can choose if he wishes to terminate the contract under the general law.

12.4   Summary

Determination by employer

The employer may determine the contractor's employment if:
o   Without reasonable cause the contractor fails to proceed diligently.
o   Without reasonable cause the contractor stops work completely.
o   The contractor becomes insolvent.

The employer may not exercise this right unreasonably or vexatiously. The procedure laid down must be followed exactly.
You must advise the employer.

Determination by contractor

The contractor may determine his own employment if:
o   The employer does not pay on time.
o   The employer or anyone for whom he is legally responsible interferes with progress of the works.

o   The employer or anyone for whom he is responsible fails to give the contractor possession of the site at the right time.

o   The employer suspends the works for one month.

o   The employer becomes insolvent.

The contractor must follow the procedure precisely.

If the contractor validly determines his own employment he gets paid:

o   The balance due to him for work begun and executed.

o   Materials on site.

o   Costs of removal from site.

The common law rights of both parties are preserved.

Chapter 13

Arbitration

13.1 General

In common with all the JCT contracts, MW 80 contains a provision for the settlement of disputes by arbitration. Arbitration is unpopular in some quarters but, since the decision of the Court of Appeal in *Northern Regional Health Authority* v *Derek Crouch Construction Co Ltd* (1984) 26 BLR 1, a defendant who is party to a contract containing a JCT arbitration clause is effectively given a right to veto litigation.

The *Crouch* case involved a contract in JCT 63 form, and the Court of Appeal held, *obiter* but unanimously and very clearly, that the High Court cannot exercise the powers of an arbitrator to 'open up, review, and revise' your certificates, decisions, and opinions. In a series of subsequent cases it has been made clear that the decision is *not* confined to those contracts where the clause is substantially similar to JCT 63, clause 35 (3).

Indeed, there has been a decision based on the wording of MW 80, article 4. This is the case of *Oram Builders Ltd* v *M. J. Pemberton & C. Pemberton* (1985) 2 Con LR 94, where the judge ruled that 'where there is an arbitration clause in general terms referring any dispute or difference between the parties concerning the contract to the arbitrator . . . then even on a narrow interpretation of the reasoning of the Court of Appeal in *Crouch*, the High Court has no jurisdiction to go behind a certificate of an architect'.

In the majority of disputes arising under MW 80, the contractor will wish to go behind your certificates and opinions, which is what really matters, and in light of the legal precedents, arbitration is going to be the norm.

Article 4 is the arbitration agreement. Under it, *any* dispute or difference concerning the contract between the employer (or you on his behalf) must be submitted to arbitration and this extends to disputes arising under the statutory tax deduction scheme unless statute provides for some other method of resolving the particular point at issue.

Arbitration is the last resort, and you should do everything possible to avoid it. Most disputes can and should be settled by negotiation, although some contractors use the existence of the arbitration clause as a lever and threaten arbitration over trivial matters.

Should this happen, it is essential that you grasp the nettle. The eventual outcome of arbitration is always uncertain, and depending on your knowledge of the contractor and the facts of the case, there are two possible approaches. A letter along the lines of Fig 13.1 may be useful where you are confident that your decision is correct, but in the vast majority of cases it is best to arrange a meeting in an endeavour to compromise. In that case, a letter based on Fig 13.2 may be helpful.

In essence, arbitration consists of the disputants referring their difference to a third party to be resolved. The essential feature of arbitration – in contrast to conciliation – is that the parties agree to be bound by the third party's decision.

Arbitration has some advantages over litigation. The arbitrator is appointed by agreement – or failing agreement by an agreed third party – and this being the case he must act as the parties require, subject to law. Consequently, important matters such as the timetable of the proceedings and the venue are fixed by agreement to suit the convenience of the parties or their witnesses and the hearing is held in private. These advantages are not shared by litigation.

Arbitration may also be marginally quicker than litigation. But it is generally just as expensive, and the speedy conduct of the proceedings depends largely on the co-operation of the parties or in finding an exceptionally robust arbitrator. Arbitration is possible on a 'documents only' basis, and this method may well be suited to many disputes arising under MW 80.

13.2 Appointing an arbitrator

As in the case of IFC 84, arbitration is possible during the currency of the contract. There is no need to wait until the works are completed or the contractor's employment has been brought to an end.

All that is needed is for there to be 'any dispute or difference concerning this Contract'. Either party can then set the procedure in motion. The first stage in that procedure is for one party to write to the other asking him to concur in the appointment of an arbitrator.

In most cases, it is the contractor who will take the initiative, but if the employer wishes to refer a dispute or difference to arbitration, you should draft a suitable letter on his behalf (Fig 13.3). It is always best to suggest the names of three people, any of whom would be acceptable as an arbitrator, but it is essential that they be independent and have no connection with the parties. It would not do, for example, to suggest your partner!

Fig 13.1

Letter from architect to contractor: to avoid arbitration

REGISTERED POST/RECORDED DELIVERY

PROJECT TITLE

Dear Sirs

Thank you for your letter dated [*insert date*] [*if appropriate, add "which has been passed to me by the employer"*].

The step you propose to take is a serious one and inevitably both parties will incur substantial costs whatever the eventual outcome. I firmly hold that my decision [*or as appropriate*] is correct and would be upheld by the arbitrator, and I have so advised the employer.

The next step is up to you. However, before proceeding further, you may wish to meet me at my office in order to discuss the matter on a reasonable basis. If you wish to do this, please telephone me to arrange a convenient appointment.

Yours faithfully

Copy: Employer

Fig 13.2
Letter from architect to contractor: where arbitration
threatened over a small matter

```
PROJECT TITLE

Dear Sir(s)

I am in receipt of your letter of [date] in which you
say that you intend to go to arbitration unless I
[specify what contractor has required you to do].
[If contractor's letter was addressed to the employer
adapt appropriately].

As I am sure you appreciate, arbitration is both
costly and time-consuming, whatever the eventual
outcome.  In the present case, the legal and related
costs are likely to outweigh any financial advantage
many times over.

While I am, of course, confident that my decision
would be upheld by an arbitrator, I am sure that upon
reflection you will agree that such a drastic course
should be avoided if at all possible.  In the
circumstances, I hope that you will agree that the
best course is for us to meet as soon as possible in
an attempt to resolve this problem.  Perhaps you will
be good enough to telephone me tomorrow.

Yours faithfully


Copy: Employer
```

Fig 13.3

Letter from employer to contractor: requesting concurrence in the appointment of an arbitrator

```
REGISTERED POST/RECORDED DELIVERY


PROJECT TITLE


Dear Sir


I/We hereby give you notice that we require the
undermentioned dispute(s) or difference(s) between us
to be referred to arbitration in accordance with
Article 4 of the contract between us dated [insert
date].  Please treat this as a request to concur in
the appointment of an arbitrator under Article 4.


The dispute(s) or difference(s) is/are:


[Specify]


I/We propose the following three gentlemen for your
consideration and require your concurrence in the
appointment of one of them within 14 days of the date
of service of this letter, failing which I/we shall
apply to the President of the Royal Institute of
British Architects/Royal Institution of Chartered
Surveyors for the appointment of an arbitrator under
Article 4.


The names and addresses of the gentlemen we propose
are:


(1) Sir Bertram Twitchett, RIBA, FCIArb,
    Fawlty Towers, Probity, Hants.


    etc.


Yours faithfully
```

The Chartered Institute of Arbitrators, 75 Cannon Street, London EC4N 5BH (Telephone 01-236-8761) is always willing to suggest the names of suitable people on request, but in order to enable the right people to be suggested, you will need to give information to the Institute about the nature of the dispute, the amount involved and so on.

It may well be that the contractor will write back to the employer suggesting other names, and the employer should not reject these out of hand. Agreeing on an arbitrator is far better than leaving the appointment in the hands of a third party.

The arbitrator must be:

o   Independent.

o   Impartial – he must not have any existing relationships with you, the employer, the contractor or anyone else involved.

o   Technically and/or legally qualified, as appropriate.

In practice, of course, if arbitration is requested, relationships between those involved have gone sour and requests to concur in the appointment of an arbitrator are often ignored. The contract therefore provides that, if the parties fail to agree on the appointment of an arbitrator within 14 days of the written request, the arbitrator will be appointed by a third party – the President or a Vice-President of the Royal Institute of British Architects or of the Royal Institution of Chartered Surveyors. Article 4 should have been completed so as to delete one or other of these appointing bodies.

Both the RIBA and RICS maintain panels of suitably qualified people from whom the appointment is made. In practice, all of them are now on the specialist panels of the Chartered Institute of Arbitrators.

Arbitrators do not work for nothing; they are professional men and they charge professional fees. Although there is no fixed scale of fees, most arbitrators charge between £350 and £1000 a day, the actual fee charged being based on the standing and expertise of the arbitrator and the complexity of the dispute. The disadvantages of a third party appointment are:

o   The appointment, once made, is binding on both parties, and the person appointed is almost impossible to remove.

o   The panel arbitrators are often very busy people and it may be difficult to get a speedy hearing.

Fig 13.4 is the sort of letter which the employer might send to an appointing body requesting an appointment.

However, it will also be necessary to complete special forms and to pay a fee to the appointing body. The RIBA form requires the applicant to undertake:

o   To provide adequate security for the due payment of the fees and expenses of the arbitrator if he so requires. [Most arbitrators do so require!]

o   To pay the arbitrator's fees and expenses whether the arbitration reaches a hearing or not. [Many arbitrators, sensibly, make it a condition

Fig 13.4
Letter from architect to appointing body

```
The Legal Secretary,          The Appointments Secretary,
Royal Institute of                       (Arbitrations),
   British Architects,           Royal Institution of
66 Portland Place,               Chartered Surveyors,
London W1N 4AD                    Great George Street,
                                          London SW1



Dear Sir

I am acting as architect for [name of employer] under
a contract in MW 80 Form, Article 4 of which makes
provision for your President to appoint an arbitrator
in default of agreement.

Will you please send me the appropriate form of
application and supporting documentation, together
with a note of the current fees payable on
application.

Yours faithfully
```

Flowchart 13.5
Arbitration

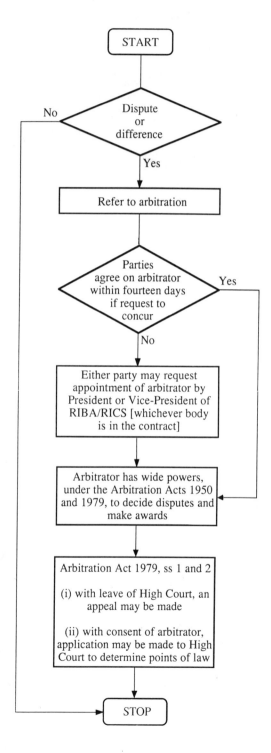

START

Dispute
or
difference — No

Yes

Refer to arbitration

Parties
agree on arbitrator
within fourteen days
if request to
concur — Yes

No

Either party may request
appointment of arbitrator by
President or Vice-President of
RIBA/RICS [whichever body
is in the contract]

Arbitrator has wide powers,
under the Arbitration Acts 1950
and 1979, to decide disputes and
make awards

Arbitration Act 1979, ss 1 and 2

(i) with leave of High Court, an
   appeal may be made

(ii) with consent of arbitrator,
application may be made to High
Court to determine points of law

STOP

of their appointment that cancellation charges are payable on a sliding scale.]

o   To take up the Award within ten days of receipt of notice of its publication by the arbitrator.

13.3   Procedure

Once the appointment is made, and accepted by the arbitrator, he will contact both parties with a view to arranging a preliminary meeting. The purpose of this meeting is to clear the ground before the actual hearing of the case and to discuss and agree a timetable and so on.

If the dispute has reached this stage, we suggest that the case should be handed over to a solicitor who is expert in arbitration law and procedure, although it is always possible for the parties to conduct their own case.

Article 4 does not define the arbitrator's powers; he derives those powers from the arbitration agreement itself and from the Arbitration Acts 1950 and 1979.

It is important to understand that, under section 1 of the 1950 Act, an arbitrator's authority is irrevocable except by leave of the High Court.

Various important powers are conferred on the arbitrator by the 1950 Act. The main ones are:

o   To order the parties to appear before him and produce all relevant documents: s. 12.

o   To administer oaths or to take affirmations: s. 12.

o   To make an Award at any time: s. 13

o   To make an interim award: s. 14.

o   To award costs: s. 18.

o   To award interest: s. 20.

Section 5 of the 1979 Act enables the arbitrator to apply to the High Court for other powers, and is now starting to be used by arbitrators to deal with reluctant parties.

Flowchart 13.5 shows the arbitration procedure.

13.4   Summary

o   MW Article 4 states that arbitration is to be used to settle disputes under the contract.

o   Arbitration can be commenced by either party at any time.

o   Failing agreement between employer and contractor on a suitable arbitrator, the appointment is made by the President or a Vice-President of RIBA or RICS.

o   The arbitrator has very wide powers.

o   Recent case law establishes an effective veto to litigation where there is an arbitration clause in the contract.

o   Avoid both litigation and arbitration if possible.

Clause Number Index

JCT Intermediate Form of Contract: An Architect's Guide

David Chappell and Vincent Powell-Smith

The IFC was welcomed by those who found the longer JCT80 unwieldy for use on contracts larger than those covered by the Minor Works form. Certainly, every architect who has not already done so will need to become familiar with it. Clearly written by two experienced authors, this guide is primarily aimed at the architect: it follows the contract through every stage, using a wealth of explanatory material such as sample letters, action flow charts and comparative tables.

Contents
The purpose and use of IFC84
Contracts compared
Contract documents and insurance
The architect's authority and duties
The contractor's obligations
The employer's powers, duties and rights
The clerk of works
Subcontractors and suppliers
Possession, practical completion, and defects liability
Claims
Payment
Determination
Arbitration
Appendix A: Form of tender and agreement NAM/T
Appendix B: NAM/SC subcontract conditions

234 × 148 mm   *232 pp*   *ISBN 0–85139–885–5*   *cloth*

Small Works Contract Documentation Third Edition

Jack Bowyer

Small contracts may still be complex, and actually differ from large ones in their administrative arrangements. There is therefore a continuing need for this book, which is devoted specifically to contractual procedure at this level. This revised and updated edition of a long-established best-seller gives advice and guidance, right through the process from inception to final account. Relevant forms are illustrated, to clarify the instructions given on how to complete them.

234 × 148 mm   164 pp   ISBN 0–85139–976–2   cloth

Professional Liability

Ray Cecil

Probably no topic causes an architect more concern than his professional liability. This book has been written to 'advise, guide and horribly warn' all architects of the changing nature of this burden. By a combination of experienced professional advice and real-life example, accompanied by practical checklists, the reader is shown what to do to avoid or minimise trouble. The author goes on to discuss the injustices of the present laws and suggests how they should be improved.

For this second edition the author has revised and updated the text and added an entirely new chapter discussing the latest developments in the field of liability, insurance and the law. There are also two new appendices, which cover the results of the RIBA's liability survey (indicating key areas of vulnerability), and the Atkins Report on project guarantee insurance (which will remain a subject of vigorous discussion for some time to come).

Contents
An outline of the law
The main areas of risk
Minimising the impact
Something has to change

'What the practitioner has been waiting for: a book written by an experienced practising architect and one who, by his own account, has been through the fire. It is well researched, and written in a highly readable style . . . I have no hesitation in saying that every practising architect should have a copy of this book.'

The Architects' Journal

234 × 148 mm   194 pp   Second edition   ISBN 0–85139–952–5
cloth

Building Contract Dictionary

Vincent Powell-Smith and David Chappell

The *Building Contract Dictionary* is the ultimate reference book for all those concerned with contract administration, from architects and quantity surveyors to contractors and solicitors. It provides authoritative answers to these and many other questions which are troublesome in practice. Its clarity of style and lack of pomposity will also make it immediately welcome to the layman—to clients who want to know what their solicitors are talking about—and to students in the building industry. It defines and explains in detail not only those words and phrases which might cause difficulty in connection with building contracts, but also the concepts encountered in relation to contracts, such as 'standard of care' or 'foreseeability'.

Within the definitions, the *Dictionary* also refers constantly to relevant legal cases; these cases not only illustrate more clearly the definitions in question, but they also provide suitable quotes from the judgments themselves

234 × 148 mm   464 pp   ISBN 0–85139–758–1   cloth

The Architect's Guide to Running a Job

Ronald Green

The architect today is greatly concerned with the ins and outs of case law, insurance and contract documents. Behind all these, however, lies something less changeable: the need for basically sound architectural practice. This is the essential structure on which any additional precautions always have to be supported.

Basic practice is the concern of this book, and the appearance of this fourth revised edition confirms the continuing need to attend to the subject. An architect more than ever has to be administrator as well as designer, and smooth economical administration will provide the conditions under which client relations can be constructive and good design can be achieved. Flow charts show a step-by-step guide to making the right moves in the right sequence, so that no single step will confuse those that follow. Within this optimal pattern, the architectural practice will find itself most free to work creatively.

Ronald Green writes from experience, for both students who want a clear guide to essentials, and professionals who want a quick reminder at their elbow. As Sir Hugh Casson says in his Foreword, 'Here you will find, set out in correct sequence – from the brisk ice-cold commonsense of its opening paragraph to the final warm-hearted words of parting advice – and without recourse to management jargon – all the many operations from site inspection to the nomination of sub-contractors, that may be met with in running a job of any size from any office, large or small. As a reference book, information chart or check-list, it will be found invaluable.'

234 × 148 mm   168 pp   illus   Fourth edition   ISBN 0–85139–061–7
cloth

Architectural Press Management Guide 1

Enquiry Form

Their Word is Law

The decisions of the Official Referees are binding on the construction
industry. But they are regularly and comprehensively reported in only one
place: in CONSTRUCTION LAW REPORTS. Whether your principal
concern is current awareness or the possession of a reliable reference shelf,
you must have every copy of the CONSTRUCTION LAW REPORTS:
the missing volume could cost you dear!
Buy peace of mind and save money by taking out an annual subscription.
And, if you have not bought the initial volumes, backdate it to volume 1,
July 1985.

Return this form, or a photocopy, to:

The Architectural Press
9 Queen Anne's Gate
London SW1H 9BY
(tel. 01 222 4333)

I wish to open a subscription to Construction Law
Reports, starting with volume . . . . . . Please send me a pro-
forma invoice/Please charge to my account no:

Name
Address

Occupation